THE BRIDGE TO ORGANIC CHEMISTRY

THE BRIDGE TO ORGANIC CHEMISTRY

Concepts and Nomenclature

CLAUDE H. YODER
PHYLLIS A. LEBER
MARCUS W. THOMSEN
Franklin and Marshall College
Lancaster, PA

A JOHN WILEY & SONS, INC., PUBLICATION

Copyright © 2010 by John Wiley & Sons, Inc. All rights reserved.

Published by John Wiley & Sons, Inc., Hoboken, New Jersey
Published simultaneously in Canada

No part of this publication may be reproduced, stored in a retrieval system, or transmitted in any form or by any means, electronic, mechanical, photocopying, recording, scanning, or otherwise, except as permitted under Section 107 or 108 of the 1976 United States Copyright Act, without either the prior written permission of the Publisher, or authorization through payment of the appropriate per-copy fee to the Copyright Clearance Center, Inc., 222 Rosewood Drive, Danvers, MA 01923, 978-750-8400, fax 978-750-4470, or on the web at www.copyright.com. Requests to the Publisher for permission should be addressed to the Permissions Department, John Wiley & Sons, Inc., 111 River Street, Hoboken, NJ 07030, 201-748-6011, fax 201-748-6008, or online at http://www.wiley.com/go/permission.

Limit of Liability/Disclaimer of Warranty: While the publisher and author have used their best efforts in preparing this book, they make no representations or warranties with respect to the accuracy or completeness of the contents of this book and specifically disclaim any implied warranties of merchantability or fitness for a particular purpose. No warranty may be created or extended by sales representatives or written sales materials. The advice and strategies contained herein may not be suitable for your situation. You should consult with a professional where appropriate. Neither the publisher nor author shall be liable for any loss of profit or any other commercial damages, including but not limited to special, incidental, consequential, or other damages.

For general information on our other products and services or for technical support, please contact our Customer Care Department within the United States at 877-762-2974, outside the United States at 317-572-3993 or fax 317-572-4002.

Wiley also publishes its books in a variety of electronic formats. Some content that appears in print may not be available in electronic formats. For more information about Wiley products, visit our web site at www.wiley.com.

Library of Congress Cataloging-in-Publication Data:

Yoder, Claude H., 1940–
   The bridge to organic chemistry : concepts and nomenclature / Claude H. Yoder, Phyllis A. Leber, Marcus W. Thomsen.
      p. cm.
   Includes bibliographical references and index.
   ISBN 978-0-470-52676-7 (Cloth : alk. paper)
   1. Chemistry, Organic. 2. Chemistry, Physical and theoretical. I. Leber, Phyllis A., 1949– II. Thomsen, Marcus W., 1955– III. Title.
     QD251.3.Y63 2010
     547–dc22

2010001897

CONTENTS

Preface		vii
1.	**Composition**	1
	Percent Composition	1
	Molecular Formula	2
	Structural Formula	2
	3D Structural Formulas	5
	Line Formulas	5
2.	**Nomenclature**	9
	Hydrocarbons and Related Compounds	9
	Alkanes	10
	Hydrocarbon Substituents	11
	Other Substituents	14
	Branched Hydrocarbon Substituents	14
	Alkanyl Names	15
	Cycloalkanes	16
	Alkenes	16
	Alkene Geometric Isomers	17
	Alkenes as Substituents	18
	Alkynes	18
	Aromatic Compounds or Arenes	19
	Functional Groups	21
	Alcohols	21
	Phenols	23
	Ethers	23
	Ketones and Aldehydes	24
	Carboxylic Acids	26
	Acid Derivatives	27
	Esters	27
	Acid Anhydrides, Halides, Amides, and Nitriles	28
	Amines	29
	Cumulative Nomenclature Problems	30
3.	**Bonding**	33
	The Lewis Model	33
	Resonance	37

	Formal Charge	38
	Generating Resonance Structures by Using Electron Flow ("Pushing Electrons")	40
	Exceptions to the Octet Rule	43
	The Valence Bond Model	44
	Triatomic Molecules	45
	Orbital Hybridization	46
	The Valence Shell Electron Pair Repulsion Model	49
4.	**Structure, Isomerism, and Stereochemistry**	**51**
	Structural Isomers	51
	Stereoisomerism	55
	Geometric Isomers	55
	Optical Isomerism	57
	Absolute Configuration	60
	Fischer Projections	62
5.	**Chemical Reactivity**	**63**
	Rate versus Extent of Reaction	63
	Mechanism	65
	Rate of Reaction	67
	Concentration of the Reactants	68
	Effect of Temperature on Rate: The Arrhenius Equation	68
	Determination of Rate Laws	69
	The Extent of Reaction: Thermodynamics	71
	Calculation of $\Delta H°$	72
	Enthalpy and Gibbs Energy of Formation	73
	Use of Bond Energies	73
	Types of Reaction	74
	Brønsted–Lowry or Proton Transfer Reactions	74
	Effect of Structure on Acidity and Basicity	75
	Proton Transfer Reactions in Organic Chemistry	79
	Electron-Sharing or Lewis Acid–Base Reactions	80
	Nucleophiles and Electrophiles	83
6.	**Reaction Mechanisms**	**87**
	Reaction Types	87
	Bond Cleavage Types	88
	Mechanism of Hydrogen–Chlorine Reaction	88
	Chlorination of Methane: A Radical Mechanism	89
	Reaction of Methyl Chloride with Hydroxide	92
	Reaction as an Ionic (Polar) Mechanism	92
	Reaction as a One-Step Mechanism	93
	Stereochemistry	93
	Effect of Leaving Group	95
	Energetics, the Reaction Profile	96
	Effect of the Solvent	97
	The S_N2 Mechanism	98
	Reaction as a Mechanism with a Trigonal Bipyramidal Intermediate	98
	Reaction of *tert*-Butyl Chloride with Water: A Two-Step Ionic Mechanism	99
Index		**103**

PREFACE

Organic chemistry is conceptually very organized and logical, primarily as a result of the mechanistic approach adopted by virtually all authors of modern organic textbooks. It continues, however, to present difficulties for many students. We believe that these difficulties stem from two major sources. The first is the need for constant, everyday study of lecture notes and textbook with paper and pencil in hand. The second is related to the integrated, hierarchical nature of organic chemistry. Many students become quickly lost simply because their knowledge of bonding, structure, and reactivity from their first course in chemistry is weak or simply forgotten. Concepts such as structural isomerism, Lewis formulas, hybridization, and resonance are generally a part of the first-year curriculum and play a very important role in modern organic chemistry. Organic nomenclature must be quickly mastered along with the critical skill of "electron pushing."

The objective of this short text is to help students review important concepts from the introductory chemistry course or to learn them for the first time. Whenever possible, these concepts are cast within the context of organic chemistry. We attempt to introduce electron pushing early and use it throughout. Nomenclature is treated in some detail, but divided into sections so that instructors can easily indicate portions they deem to be most important. In the last chapter we provide an introduction to mechanisms that utilizes many of the concepts introduced earlier—Lewis acid–base chemistry, rate laws, enthalpy changes, bond energies and electronegativities, substituent effects, structure, stereochemistry, and, of course, the visualization of electron flow through the electron-pushing model. Hence, the chapter shows the value of certain types of reasoning and concepts and contains analyses not commonly found in organic texts.

The text is designed for study either early in the organic course or, preferably, prior to the beginning of the course as a bridge between the introductory course and the organic course. Because the text is designed to be interactive, it is essential that the student study each question carefully, preferably with the answer covered to thwart the ever-present tendency to "peek." After careful consideration of each question using pen and paper, the answer can then be viewed and studied. In this bridge between introductory and organic chemistry we have made a serious effort to review topics as the reader progresses through the text and to focus on important concepts rather than simply to expose the student to different types of organic reactions.

The authors are indebted to Dr. Ronald Hess (Ursinus College), Dr. David Horn (Goucher College), Dr. Anne Reeve (Messiah College), Dr. Edward Fenlon (Franklin and Marshall College), Audrey Stokes, Brittney Graff, Victoria Weidner, Chelsea Kauffman, Mallory Gordon, Allison Griffith, and William Hancock-Cerutti for helpful suggestions.

CLAUDE H. YODER
PHYLLIS A. LEBER
MARCUS W. THOMSEN

1

COMPOSITION

Carbon forms a vast variety of covalent compounds, many of which occur naturally in biological systems. Besides their importance to plant and animal life, these compounds offer examples of a wide array of structures that challenge chemists to synthesize them. Most carbon compounds are composed of only a small number of elements: carbon, hydrogen, oxygen, nitrogen, and the halogens.

When a chemist prepares or encounters a new substance, the first question that arises is "What elements are present?" After this is determined, the questions become increasingly sophisticated: What is the weight ratio of the elements? How are the atoms of the elements bonded to one another? What is the geometric arrangement of the atoms? For organic compounds, in which the elements are generally attached by covalent bonds to form molecules, the chemist ultimately would like to know the three-dimensional (3D) shape of the molecule. This shape, or structure, can determine how the molecule reacts with various reagents and can also affect physical properties such as boiling point and density. In the following section we progress from the question of the weight ratio of the elements to a series of formulas that reveal different aspects of the structure of molecules. Our goal is to produce a formula that expresses the shape of the entire molecule.

PERCENT COMPOSITION

The simplest way to express the composition of a compound is the mass percentage of its constituent elements. Let's make sure that you remember how to convert percent composition to the empirical formula.

Q The organic compound benzene contains 92.3% carbon and 7.7% hydrogen. Calculate the empirical formula of benzene.

A Probably the simplest way to proceed is to assume that you have a sample of benzene that weighs 100 g. In 100 g of benzene there are

$$0.923 \times 100 \text{ g} = 92.3 \text{ g of carbon}$$

$$0.077 \times 100 \text{ g} = 7.7 \text{ g of hydrogen}$$

The *empirical formula* presents the simplest whole-number ratio of the number of moles of each element in the compound. The number of moles of each element is easily obtained by dividing by the atomic weight of each element.

$$92.3 \text{ g C}/12.01 \text{ g/mol} = 7.69 \text{ mol C}$$

$$7.7 \text{ g H}/1.008 \text{ g/mol} = 7.6 \text{ mol H}$$

These numbers are the same within experimental uncertainty; hence, the ratio of the number of moles of carbon to that of hydrogen is one to one. The formula CH therefore represents the simplest whole-number ratio of the number of moles of carbon to the number of moles of hydrogen. ∎

The Bridge to Organic Chemistry: Concepts and Nomenclature
By Claude H. Yoder, Phyllis A. Leber, and Marcus W. Thomsen
Copyright © 2010 John Wiley & Sons, Inc.

2 COMPOSITION

Benzene, like most organic compounds, is a molecular covalent compound.

Q How would you know that benzene is a covalent compound, rather than an ionic compound?

A In general, the bonding between two elements becomes more ionic as the difference in electronegativity of the elements increases. In $CaCl_2$ the difference is so large [3.0 (Cl) – 1.0 (Ca) = 2.0] that the calcium is present as the +2 cation and chlorine as the –1 anion. For methane, on the other hand, the difference in electronegativity is small [2.5 (C) – 2.1 (H) = 0.4] and the electrons are shared within a covalent bond. For a compound that contains only carbon and hydrogen, such as benzene, we can reasonably assume that the bonding is covalent. The majority of organic compounds that you will study are *molecular*; that is, the atoms are held together by covalent bonds within a molecule. ■

We now need to determine how many atoms of each element are present in one molecule of benzene. You may be thinking that if CH is the simplest ratio of atoms in the compound, then each molecule should contain one carbon and one hydrogen atom. However, the empirical formula does not tell us how many atoms of each element are present in each molecule. For example, there could be two atoms of carbon and two atoms of hydrogen, or three and three, and so on, in one molecule.

MOLECULAR FORMULA

In order to determine the molecular formula from the empirical formula, we need to know the molecular mass (molecular weight). This value is the mass of one mole of molecules and can be determined experimentally by a number of methods, including mass spectrometry. For benzene the molecular weight is 78 g/mol.

Q How can you use the molecular weight to convert the empirical formula to the molecular formula?

A Assume for a minute that CH is the molecular formula; the molecular weight would then be 12.01 + 1.008 = 13.02 g/mol. If we divide the molecular weight of 78 g/mol by the molar mass of the unit CH

$$(7.8 \text{ g/mol})/(13 \text{ g/mol}) = 6$$

we find that there are six "CH" units in each molecule of benzene. The molecular formula may be written as $(CH)_6$, but it is customary to write it as C_6H_6. ■

STRUCTURAL FORMULA

The next step in determining the structure of a compound is to determine how the atoms are arranged or attached to one another. Now that we know that a molecule of benzene has six carbons and six hydrogen atoms, we can speculate about some ways in which these atoms can be arranged.

Q Can you think of a simple way to arrange six carbons and six hydrogens in a line?

A One arrangement of these atoms is as follows:

$$\text{H}\cdots\text{C}\cdots\text{H}\cdots\text{C}\cdots\text{H}\cdots\text{C}\cdots\text{H}\cdots\text{C}\cdots\text{H}\cdots\text{C}\cdots\text{H}\cdots\text{C}$$
$$(1)$$
■

This sequence of atoms represents the connectivity of atoms; that is, the specific way that atoms are connected to one another.

Statement. The structural formula expresses the connectivity within a molecule.

You should remember that normally hydrogen does not form more than one covalent bond, so arrangement (**1**) is not very likely. You could imagine groupings of hydrogen atoms around atoms such as

$$\begin{array}{ccccccc} & \text{H} & & & & & \text{H} \\ \text{H}\cdots\text{C}\cdots\text{C}\cdots\text{C}\cdots\text{C}\cdots\text{C}\cdots\text{C}\cdots\text{H} \\ & \text{H} & & & & & \text{H} \end{array}$$
$$(2)$$

Statement. In both representations above it is important to realize that the dashed lines are used to indicate attachments or connectivities of atoms.

These lines do not indicate electron-sharing bonds. Eventually, however, we will need to determine whether the atoms *could* be attached to one another by covalent bonds and that will require use of the Lewis model.

For organic compounds we use a number of models to explain covalent bonding, one of the most important of which is the Lewis (electron dot) model.

Statement. Good Lewis structures usually involve four bonds at carbon, three to nitrogen, two to oxygen, and one to hydrogen or a halogen.

Of course, Lewis structures must contain the appropriate number of electrons and, where possible, must obey the octet rule (for hydrogen, only two electrons). In order to determine whether either structure **1** or **2** *might* be a reasonable structure, we should see if we can write a conventional Lewis structure for each.

Q Write a Lewis structure for structure **2**.

A For structure **2**, we could write a perfectly acceptable Lewis structure:

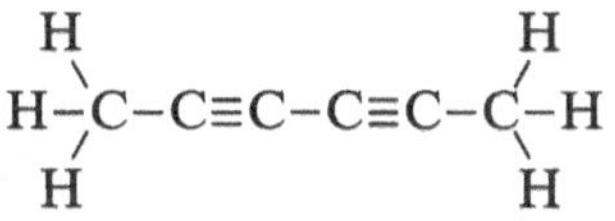

$$H-\underset{H}{\overset{H}{C}}-C\equiv C-C\equiv C-\underset{H}{\overset{H}{C}}-H$$

Although a good Lewis structure can be written for structure **2**, this does not mean that structure **2** is the correct structural formula for benzene. In order to determine whether this representation is the structural formula, we must perform either chemical or spectroscopic tests. For example, the Lewis structure for structure **2** contains both carbon–carbon triple bonds and carbon–carbon single bonds. We need a method that can tell us if these two types of bonds are present in benzene. ∎

Although chemical methods can be used to determine whether a double or triple bond is present, this determination is more commonly accomplished using spectroscopic methods. These methods, all of which involve irradiating a sample with electromagnetic radiation, include infrared (IR) and ultraviolet–visible (UV–VIS) spectroscopy, as well as nuclear magnetic resonance (NMR) spectroscopy. The colorimeter (e.g., the common Spectronic 20) that you may have used in general chemistry courses employed radiation in the visible region to change the electronic energy levels of the molecule. Infrared spectroscopy, which uses lower frequencies of "light," changes the energies of the vibrations of different groups of atoms within a molecule. You need not worry at the moment about learning about the various spectroscopic methods, but we use a few such methods below to demonstrate how the structures of molecules are determined.

Q Although we will not discuss infrared spectroscopy in any detail, it is helpful to know that different types of bonds absorb different frequencies of IR light. In general, the stronger the bond, the higher the frequency of the light required to increase the vibrational amplitude of the bond vibration. Look at the carbon–carbon bonds in structure **2** and determine whether the carbon–carbon single or triple bonds will absorb higher frequencies of infrared radiation.

A Because the triple bond is stronger than the single bond, the triple bond requires higher frequencies of radiation. Therefore structure **2** would have at least two peaks in the carbon–carbon region of its IR spectrum. However, when the infrared spectrum of benzene is examined, there is no peak due to a triple bond. Consequently, structural formula **2** is not correct for benzene. ∎

Q Write a Lewis structure for the connectivity of atoms portrayed by the following structure:

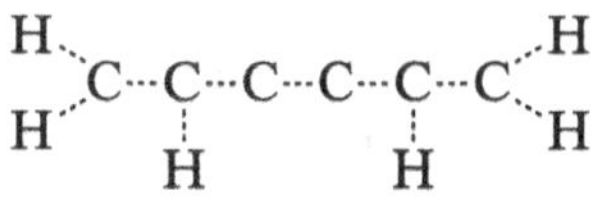

$$\underset{H}{\overset{H}{\diagdown}}C\cdots C\cdots\underset{H}{C}\cdots C\cdots\underset{H}{C}\cdots C\overset{H}{\underset{H}{\diagup}}$$

(3)

A

$$H-\underset{H}{\overset{H}{C}}=\underset{}{\overset{H}{C}}-C\equiv C-\underset{}{\overset{H}{C}}=\underset{H}{\overset{H}{C}}-H$$

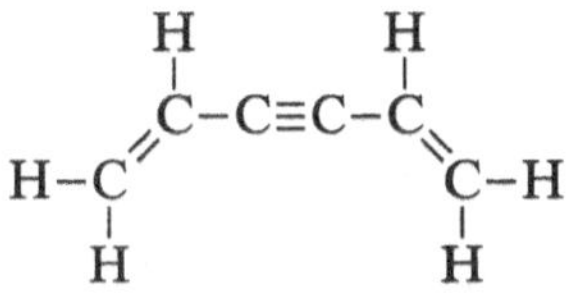

This arrangement contains carbon–carbon single, double, and triple bonds, and at least two different environments for hydrogen atoms. In order to determine whether this is a reasonable structural formula, we will use another spectroscopic technique—nuclear magnetic resonance (NMR) spectroscopy—to determine the number of chemically different carbons in a molecule. When carbons are "chemically different," they generally have different electron densities around them. The number of chemically different carbon atoms in a molecule can be determined from the symmetry of the molecule. For the structure immediately above, there is a plane of symmetry that divides the molecule in half. The plane (see next structure below) cuts through the triple bond in the center of the molecule. ∎

4 COMPOSITION

Q Examine the structure above and determine how many chemically different carbons there are. Remember that there is a plane of symmetry cutting the molecule in two halves. These two halves are mirror images of one another.

A The mirror plane makes the two terminal carbon atoms the same (see the following structure); the two C–H carbons are the same, and the two atoms of the triple bond have the same electronic environment. Thus, there are three chemically different carbon atoms. If we obtain the NMR spectrum of the carbon atoms in this structure, the spectrum would indicate three different carbons.

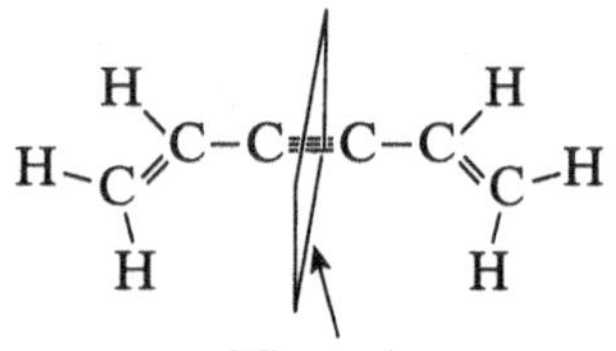

$$H-\underset{H}{\overset{H}{C}}\!\!\diagdown\!\!C-C\!\equiv\!\!C-C\!\!\diagup\!\!\underset{H}{\overset{H}{C}}-H$$

Mirror plane

Q How many different carbons are there in a molecule of oxalic acid as shown below?

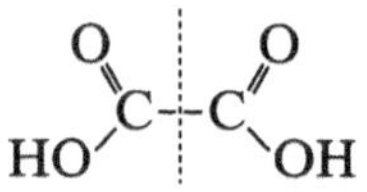

$$HO\overset{O}{\underset{}{\diagdown}}C\!-\!C\overset{O}{\underset{OH}{\diagup}}$$

A The plane of symmetry running through the carbon–carbon bond in the center of the molecule makes the two carbons equivalent. Therefore, this compound has only one type of carbon. ■

The actual NMR spectrum for the carbon atoms of benzene contains evidence for only one type of carbon in benzene, and structure **3** is not the correct structural formula for benzene.

If we continue this process of writing and testing structural formulas long enough, we will eventually arrive at one that satisfies all of the spectroscopic and chemical information. It is the structure shown below, in which the carbon atoms are at the corners of a perfect hexagon with a hydrogen attached to each of the carbons.

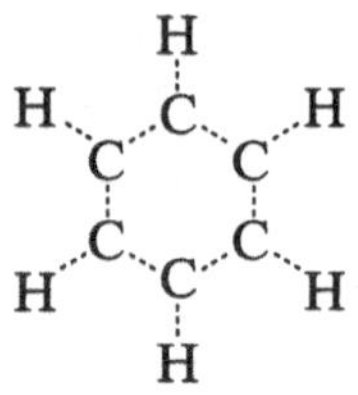

It is not obvious from this structural formula that the molecule is planar (i.e., with all atoms lying in one plane), as we will see during our discussion of three-dimensional structural formulas. This structure was first suggested in 1865 by the German chemist Friedrich August Kekule (1829–1896), who claimed that he derived the structure from a dream about a snake biting its own tail.

Now that we know how the atoms in benzene are arranged, we will learn how the electrons are arranged by writing the Lewis structure.

Q Write a Lewis structure for benzene. Notice that the molecule has a total of 30 valence electrons (4 from each carbon and 1 from each hydrogen) that must be arranged to give each atom eight electrons, except hydrogen, which must have two. If you do not remember how to write Lewis structures, rest easy because we will cover this topic in more detail in Chapter 3.

A The Lewis structure below satisfies the octet rule and has the correct number of electrons.

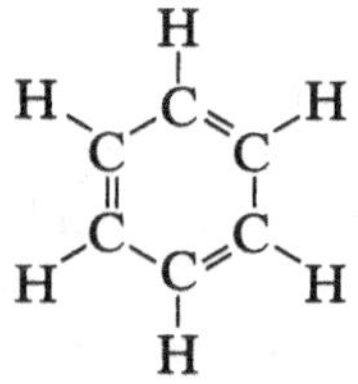

■

We will find later that this Lewis structure does not do justice to some of the properties of benzene and that it must be modified to make all carbon–carbon linkages the same. (The word *linkage* refers to the connection between two atoms. Normally, the word *bond* is used, but this also connotes a shared pair of electrons.) This modification, known as *resonance hybridization*, is shown below by writing two Lewis structures with a double-headed (double-barbed) arrow between them. The resonance hybrid of the two individual Lewis structures is a better representation of the electronic formula of benzene:

In the resonance hybrid each carbon is identical to the other carbons, and each carbon–carbon bond is the same as the other carbon–carbon bonds. Therefore, this structure is consistent with the carbon NMR spectrum.

3D Structural Formulas

Because much of the behavior of organic compounds depends on their shapes, we need to go one step farther and determine the geometry of the benzene molecule. We could speculate that the hexagonal structure of benzene could have a 3D shape like one of the following:

(a)   (b)

These diagrams are somewhat limited in their ability to portray three-dimensional structure, and we must therefore rely on some conventions to show spatial orientation.

Statement. The solid wedges indicate bonds that come out of the paper toward the reader; the dashed lines or dashed wedges indicate bonds that go behind the paper away from the reader; solid lines are used to represent bonds in the plane of the paper (or parallel to the plane of the paper).

Q Use the convention given above to draw a 3D structural formula for methane. Remember that methane has a carbon at the center of a tetrahedron with a hydrogen atom at each corner of the tetrahedron.

A The tetrahedron can be visualized as two perpendicular planes, each containing the carbon and two hydrogen atoms. You will need a model to fully appreciate this geometry. The formula below conveys these two planes quite clearly.

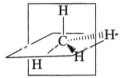

■

Let's return now to the shape of benzene. The correct shape of benzene is shown by representation **b** above. In this representation all of the carbon atoms and all of the hydrogen atoms are in the same plane. This geometry for benzene is also shown in Figure 1.1 with a ball-and-stick representation, with the atoms in the front drawn larger to give a 3D perspective.

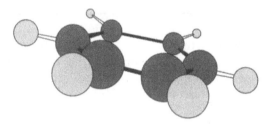

Figure 1.1. A ball-and-stick representation of benzene. The atoms closer to the reader are drawn larger to give a 3D perspective.

Line Formulas

Representation of organic molecules, many of which are large and complex, is greatly simplified by using line formulas or line structures. A *line structure* for benzene is shown below.

If the line structure is compared to our previous representations, we can see that each intersection of line segments represents a carbon. Because the line structure is a Lewis formula, *each neutral carbon atom must have sufficient hydrogen atoms surrounding it to produce an octet of electrons*. In the line structure of propanone (CH₃COCH₃), shown below, notice that the lines going to the C=O group represent methyl (CH₃) groups. In other words, carbons appear at the *intersection* of line

6 COMPOSITION

segments and at the *terminus* of a line segment unless another atom appears at those points.

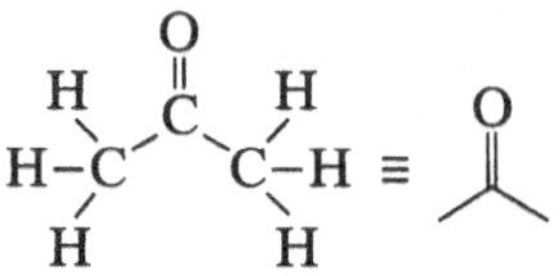

Propanone (also known as *acetone*) is another good compound to commit to memory.

Q Give the line formula for ethanol. Because ethanol is an important compound, try to remember the formula CH₃CH₂OH.

A

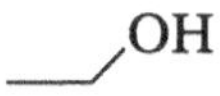

Q Draw the line formula for 1,1,1-trifluoropropanone. The nomenclature 1,1,1-trifluoro- tells you that there are three fluorines attached to one of the terminal carbons of propanone.

A

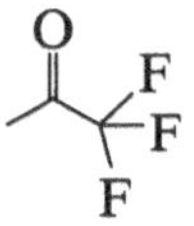

Be sure that you can also write formulas with all of the atoms "written out" and as *condensed formulas*. Here are these two types of formulas for 1,1,1-trifluoropropanone:

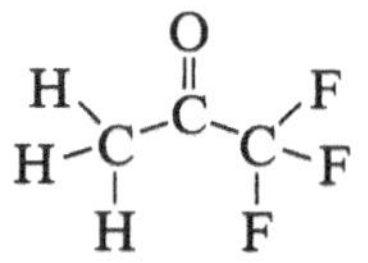

"Written out" formula   Condensed formulas

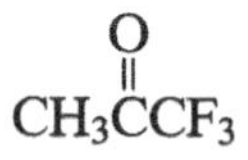

Q Condensed formulas can be written in a variety of different ways. Which one of the following formulas is not correct for 1,1,1-trifluoropropanone?

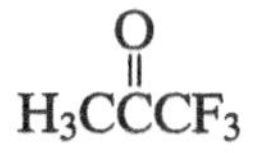

H_3CCCF_3 CH_3COF_3C
(a)   (b)

A In structure **a** the hydrogens are written before the carbon to indicate that they are attached to the carbon. This is an acceptable formula. In structure **b** the fluorines are written before the terminal carbon, but this formula is unacceptable because of the convention that attached atoms always follow the atom to which they are attached. Exceptions to this rule only occur on the left side of the formula where there can be no confusion about the meaning of either CH_3 or H_3C.

Q Provide line structures for each of the following:

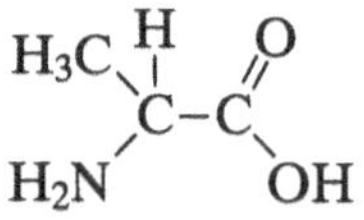

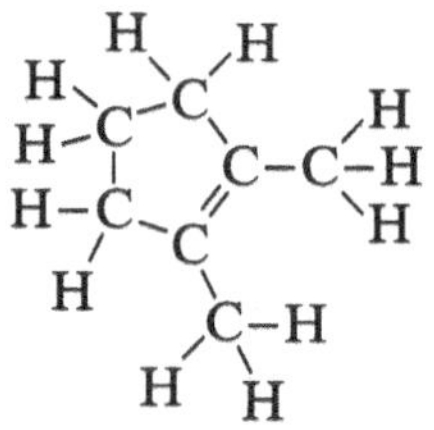

A

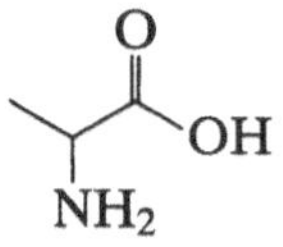

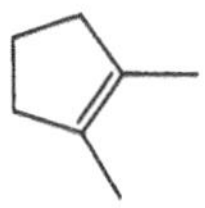

Now, look at structure **a** of the common pain reliever ibuprofen below and compare it to line structure **b**. The line structure **b** should look less cluttered and confusing to you.

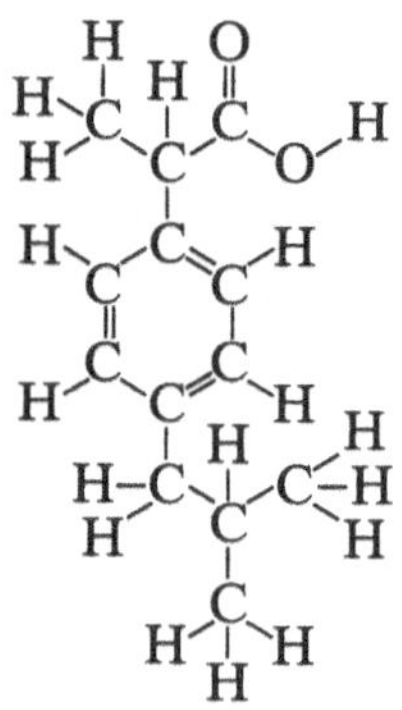

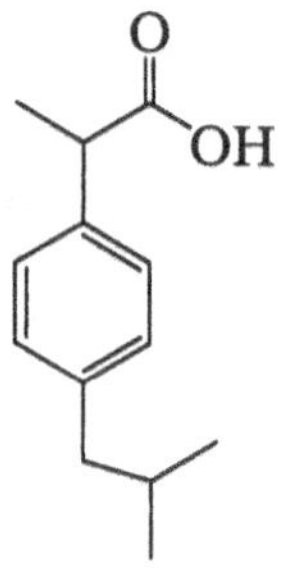

(a)   ibuprofen   (b)

Q Provide line structures for each of the following molecules:

$$\begin{array}{c} \overset{OH}{}\overset{CH_3}{} \\ CH_3CH_2CH_2CHCH_2CHCH_3 \end{array}$$

$$\begin{array}{c} \overset{Br}{} \\ CH_3CH_2CCH_2CH_2CH_2OCH_2CH_3 \\ \underset{CH_2CH_3}{} \end{array}$$

$$\begin{array}{c} \overset{CH_3}{} \\ CH_3CHO CH_2CH_2CH_3 \\ C=C \\ CH_3CH_2CH_2CH_2 CH_2CH_2CH_3 \end{array}$$

$$\begin{array}{c} Cl CH_2CH_2CH_3 \\ HC-CH \\ HC=CH \end{array}$$

A

2

NOMENCLATURE

The name of a compound must be unambiguous; that is, the name can leave no question about how to draw the structural formula of the compound. The International Union of Pure and Applied Chemists (IUPAC) has provided rules for names and periodically reviews and rewrites these rules. However, before the IUPAC committee began to provide the systematics of nomenclature, chemists named compounds using rules developed over the years, or simply through the use of some trivial name. The compound

$$H_3C-\overset{\overset{O}{\|}}{C}-CH_3$$

was at one time known only as *acetone*, because it can be obtained by heating vinegar, which was known as *acetum*; *acetone* means "daughter of acetum." Later, acetone was given the common name of *dimethyl ketone*, and then with the advent of the IUPAC rules acetone was named *propanone*. Most chemists, however, still use the trivial name *acetone*. Nevertheless, most of our discussion of nomenclature will follow the IUPAC rules, although you will also learn the common system and even some trivial names.

We start by dividing organic compounds into two major classes: hydrocarbons and compounds with functional groups. Hydrocarbons contain only carbon and hydrogen. Certain hydrogen replacements, called *functional groups*, give organic molecules characteristic chemical behaviors that are very different from those of hydrocarbons. For example, when a carbonyl group (C=O) is present in a structure, as is true for the ketone acetone, reagents that would not react with the parent hydrocarbon will react vigorously with the carbonyl group.

HYDROCARBONS AND RELATED COMPOUNDS

The simplest type of carbon compound, the hydrocarbons, contains carbon atoms linked to one another and also to hydrogen. There are four main kinds of hydrocarbons: (1) *alkanes*, in which all the carbon–carbon linkages are single bonds; (2) *alkenes*, in which one or more of the carbon–carbon linkages are double bonds; (3) *alkynes*, in which one or more of the carbon–carbon linkages are triple bonds; and (4) *aromatics*, in which the benzene ring is present. Alkenes and alkynes are sometimes referred to as *unsaturated compounds* because the linked carbon atoms are not bonded to as many hydrogen atoms as possible; that is, the carbons are not saturated with respect to hydrogen. Aromatic compounds (benzene relatives) have a special arrangement of alternating carbon–carbon double bonds, and represent a separate category of unsaturated hydrocarbons.

Q Convert each of the following compounds to its saturated analog:

$$H_2C=C\overset{CH_3}{\underset{H}{\diagdown}} \qquad H_3C-C\equiv C-CH_3$$

The Bridge to Organic Chemistry: Concepts and Nomenclature
By Claude H. Yoder, Phyllis A. Leber, and Marcus W. Thomsen
Copyright © 2010 John Wiley & Sons, Inc.

CH₃–CH₂–CH₃   CH₃–CH₂–CH₂–CH₃ ∎

Q Reason by analogy with what you did in the question above to predict the saturated analog of propanone. If you have forgotten the formula of propanone, it may be helpful to know that it contains a C=O bond.

A

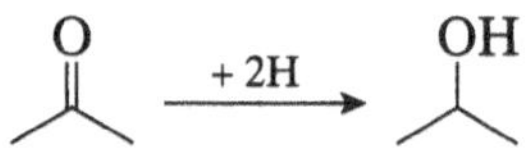

The compound with the OH group is an alcohol and is the saturated analog of the ketone propanone (acetone). ∎

Q Which of the following compounds is an alkene?

A The cyclic compound in the middle is an alkene. The compound on the left is an alkyne; the compound on the right is an alkane. ∎

Q How many carbons are there in the compound on the left in the previous question? Make sure that you can write out all the carbons and hydrogens for this compound.

A There are four carbons in the formula for the alkyne on the left.

HC≡C–CH₂–CH₃ ∎

Alkanes

Even though methane (CH₄) is considered an alkane, the simplest hydrocarbon that contains a single carbon–carbon bond is ethane (CH₃–CH₃), but alkanes exist that contain many carbons linked together. In fact, one of the very special features of the chemistry of carbon is the extent to which this linking of atoms can occur; it is at least partly responsible for the formation of very large molecules that form polymers and biologically active organic compounds. The formulas and names of some *straight-chain alkanes* (alkanes whose carbon atoms can be written on a straight line) are given in Table 2.1. All the names end in *-ane*, and from pentane to decane the names are derived from the Greek word for the number of carbon atoms in one molecule of the compound.

TABLE 2.1. Nomenclature for Straight–Chain Alkanes

Compound	Name
CH₄	methane
CH₃CH₃	ethane
CH₃CH₂CH₃	propane
CH₃(CH₂)₂CH₃	butane
CH₃(CH₂)₃CH₃	pentane
CH₃(CH₂)₄CH₃	hexane
CH₃(CH₂)₅CH₃	heptane
CH₃(CH₂)₆CH₃	octane
CH₃(CH₂)₇CH₃	nonane
CH₃(CH₂)₈CH₃	decane

Q Write out the formula of a *branched* five-carbon alkane.

A

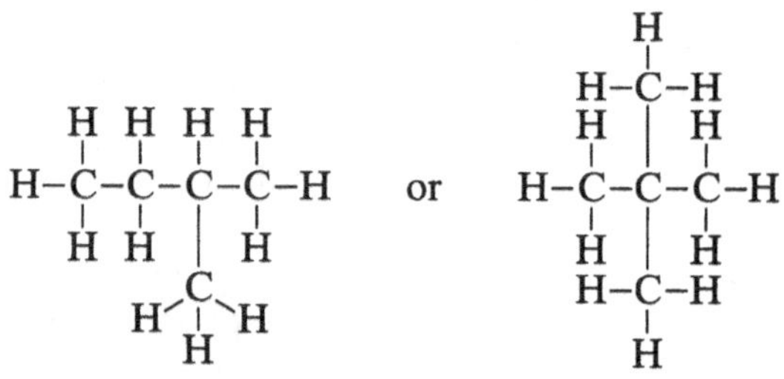

∎

Q Refer to Table 2.1 and write out the formula of pentane with all the bonds shown clearly as above. Also write the formula using a condensed formula and using a line formula.

A

```
    H H H H H
    | | | | |
H - C-C-C-C-C - H
    | | | | |
    H H H H H
```
(a)

CH₃CH₂CH₂CH₂CH₃   /\/\
(b)   (c)

As we have seen, the structural formulas of organic molecules can be written in a number of ways. The formula **a** shows clearly all of the attachments, **b** is a more condensed formula, and **c** is the line formula. Note that the line formula must be written with the lines at angles, rather than in a straight line, in order to show the line vertices (intersections). ■

Q A student drew pentane like this:

```
    H   H H H
    |   | | |
H - C - C-C-C - H
    |   | | |
    H   H H C - H
            |
            H  H
```

Is it incorrect to have the end CH₃ pointed down rather than at the end of a straight line?

A This is a perfectly good structure (although not quite as aesthetically pleasing as the representation with all the carbons in a straight line) and it is a straight-chain alkane because the carbons can be written without branching. ■

Q What is wrong with this five-carbon structure?

```
    H H H H
    | | | |
H - C-C-C-C
    | | | |
    H H | H
        C
       /|\
      H | H
        H
```

 This structure is a branched five-carbon alkane identical to the branched alkane that we looked at earlier, except that there is a problem with the structure. The carbon on the right side has only three bonds

HYDROCARBONS AND RELATED COMPOUNDS 11

TABLE 2.2. Alkyl Groups

Alkane	Alkyl Group	Alkyl Group Name
CH₄	–CH₃	methyl
CH₃CH₃	–CH₂CH₃	ethyl
CH₃CH₂CH₃	–CH₂CH₂CH₃	propyl
CH₃(CH₂)₂CH₃	–CH₂(CH₂)₂CH₃	butyl
CH₃(CH₂)₃CH₃	–CH₂(CH₂)₃CH₃	pentyl
CH₃(CH₂)₄CH₃	–CH₂(CH₂)₄CH₃	hexyl
CH₃(CH₂)₅CH₃	–CH₂(CH₂)₅CH₃	heptyl
CH₃(CH₂)₆CH₃	–CH₂(CH₂)₆CH₃	octyl
CH₃(CH₂)₇CH₃	–CH₂(CH₂)₇CH₃	nonyl
CH₃(CH₂)₈CH₃	–CH₂(CH₂)₈CH₃	decyl

to it—one to another carbon and two bonds to two hydrogen atoms. *Every carbon in an alkane requires four bonds* in order to obey the octet rule. This structure therefore does not contain the correct number of hydrogen atoms. The group on the right side should be –CH₃. ■

If you try to name the branched hydrocarbon above (after changing the –CH₂ to a –CH₃ group), you will encounter difficulty. It is not pentane, and yet it does have five carbons. It looks like butane with a CH₃ group attached to the second carbon from the end. This CH₃ group is derived from methane (CH₄), by removing one of the hydrogen atoms, and it is called the *methyl group*. In order to name this and other branched hydrocarbons we need to learn about alkyl groups.

Hydrocarbon Substituents. Many molecules contain an alkane unit less one hydrogen atom as part of their structure. These groups are named by replacing the *-ane* ending in the alkane's name by *-yl*. For example, CH₃CH₃ is ethane, and CH₃CH₂ is an ethyl group. The names of these alkyl groups are given in Table 2.2.

Q If you have not yet memorized the number of carbons in each one of the alkanes, remember that from pentane to decane, the Greek or Latin prefixes indicate the number of carbons. From methane to butane you simply need to memorize them. Give the formula for the propyl group.

A The propyl group is derived from propane (CH₃CH₂CH₃) by removing a hydrogen atom from the end of the propane. Hence, the propyl group is CH₃CH₂CH₂ and because the right-hand carbon has

12 NOMENCLATURE

only three bonds, we can attach the propyl group to a carbon in another molecule. If we want to make a branched hydrocarbon out of heptane ($CH_3CH_2CH_2CH_2CH_2CH_2CH_3$) using the propyl group, we must remove a hydrogen from one of the carbons of heptane so that we can attach the propyl group. The propyl group is referred to as a *substituent* because it substitutes for a hydrogen. There are three carbons in heptane from which we could remove a hydrogen in order to make the branched hydrocarbon. ∎

Q There are seven carbons in heptane, so why are there only three to which the propyl group can be attached?

A Let's look first at heptane with a hydrogen removed from the second carbon from the left.

$$CH_3\overset{*}{C}HCH_2CH_2CH_2CH_2CH_3$$

This alkyl group seems to be different from the structure derived by removing a hydrogen from the second carbon from the right.

$$CH_3CH_2CH_2CH_2CH_2\overset{*}{C}HCH_3$$

But, in fact, if you rotate the first one 180° to the right, you generate the second structure. This means that the two formulas are actually the same; they are just written differently. The same is true if you remove a hydrogen from the third carbon from the left (or the corresponding carbon counted from the right). ∎

Q Why not remove the hydrogen from the carbon on either end of the molecule?

A If we were to place the propyl or any other group on the end, we would simply expand the chain of carbons, rather than generate a *branched* hydrocarbon. If you add the propyl group to the end of heptane, you obtain decane.

$CH_3CH_2CH_2CH_2CH_2CH_2CH_2-\ +\ -CH_2CH_2CH_3$
$\longrightarrow CH_3CH_2CH_2CH_2CH_2CH_2CH_2CH_2CH_2CH_3$ ∎

 Draw a condensed formula for the branched hydrocarbon obtained by placing the propyl group on the fourth carbon of heptane.

A

$$CH_3CH_2CH_2CHCH_2CH_2CH_3$$
$$|$$
$$CH_2CH_2CH_3$$

The compound that you just generated is named 4-propylheptane, and the propyl group substitutes for a hydrogen on a hydrocarbon. ∎

Statement. According to the IUPAC rules, we must name the group (substituent), the hydrocarbon parent to which it is attached, and we must indicate by a number the carbon in the parent to which it is attached. Moreover, the parent hydrocarbon chain of carbons must be numbered so that the substituent receives the lowest possible number.

Q Which terminal carbon of heptane is assigned the number 1?

A In this case, we can number from either side in the structure because the fourth carbon is in the middle of the chain. In other words, the propyl group will be attached to the fourth carbon regardless of whether we start the numbering on the left or the right side of heptane. ∎

 Now generate the branched hydrocarbon derived by placing the propyl group on the second carbon of the heptane chain.

A

$$CH_3CHCH_2CH_2CH_2CH_2CH_3$$
$$|$$
$$CH_2CH_2CH_3$$

It would seem that we could name this compound 2-propylheptane, but there is another IUPAC rule that tells us that we must use the longest chain of carbons as the "parent" hydrocarbon. In order to see that the parent is now a nine-membered chain, a nonane, we could exchange the groups in the formula as shown below.

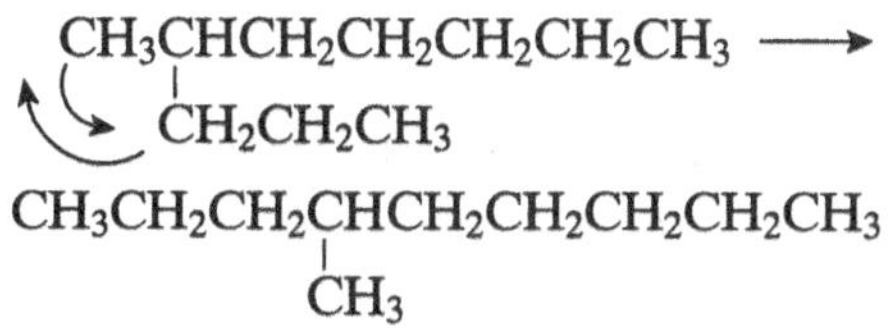

$$\underset{\underset{\text{CH}_2\text{CH}_2\text{CH}_3}{|}}{\text{CH}_3\text{CHCH}_2\text{CH}_2\text{CH}_2\text{CH}_2\text{CH}_3} \longrightarrow$$

$$\underset{\underset{\text{CH}_3}{|}}{\text{CH}_3\text{CH}_2\text{CH}_2\text{CHCH}_2\text{CH}_2\text{CH}_2\text{CH}_3}$$

Q What is the name of this hydrocarbon?

A 4-methylnonane. Notice that we had to number the nonane chain from the left side in order to give the methyl group the lowest possible number on the parent.

Q Provide the name for each of the following branched hydrocarbons:

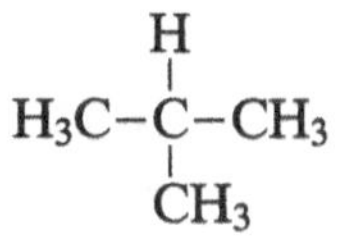

$$\underset{\underset{\text{CH}_3}{|}}{\overset{\overset{\text{H}}{|}}{\text{H}_3\text{C}-\text{C}-\text{CH}_3}} \qquad \underset{\underset{\underset{\text{CH}_3}{|}}{\underset{\text{CH}_2}{|}}}{\overset{\overset{\text{H}}{|}}{\text{H}_3\text{C}-\text{C}-\text{CH}_2\cdot\text{CH}_3}}$$

(a) 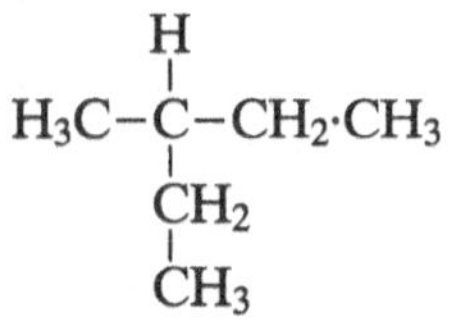 (b)

$$\underset{\underset{\text{CH}_3}{|}}{\overset{\overset{\text{H}}{|}}{\text{CH}_3-\text{CH}_2-\text{C}-\text{CH}_3}}$$

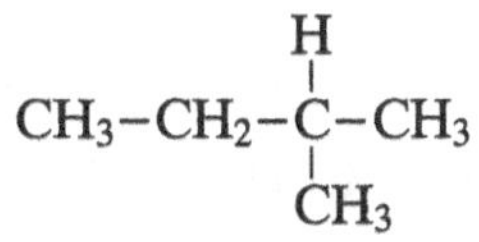

(c)

A For compound **a** the chain is numbered as follows:

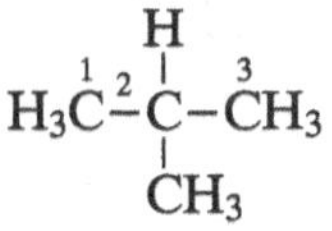

$$\underset{\underset{\text{CH}_3}{|}}{\overset{\overset{\text{H}}{|}}{\text{H}_3\overset{1}{\text{C}}\overset{2}{-}\text{C}-\overset{3}{\text{CH}}_3}}$$

For compound **b** the longest chain is shown below, both as written above and as it could be written in a straight line.

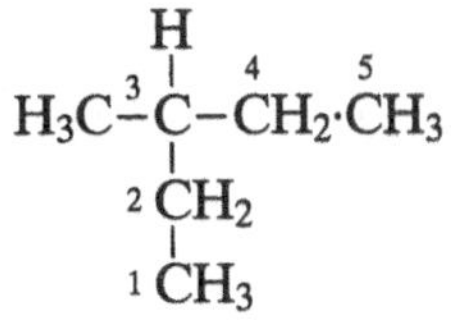

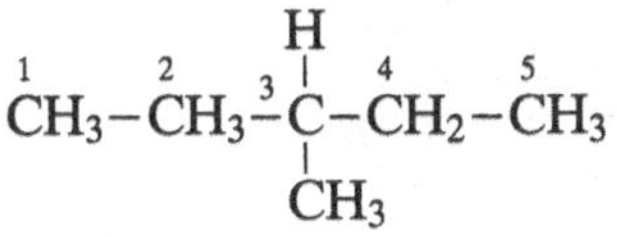

$$\underset{\underset{\underset{{}^{1}\text{CH}_3}{|}}{\underset{{}^{2}\text{CH}_2}{|}}}{\overset{\overset{\text{H}}{|}}{\text{H}_3{}^{3}\text{C}\overset{4}{-}\text{C}-\overset{5}{\text{CH}}_2\cdot\text{CH}_3}} \qquad \underset{\underset{\text{CH}_3}{|}}{\overset{\overset{\text{H}}{|}}{\overset{1}{\text{CH}}_3-\overset{2}{\text{CH}}_3\overset{3}{-}\text{C}-\overset{4}{\text{CH}}_2-\overset{5}{\text{CH}}_3}}$$

In both **a** and **b** the branch occurs midway in the longest chain, and so it is immaterial from which end the num-

bering begins. In compound **c**, the branch occurs closer to one end of the chain; thus, numbering the carbons begins from the end position closest to the branch so that the methyl group is attached to carbon 2 rather than carbon 3.

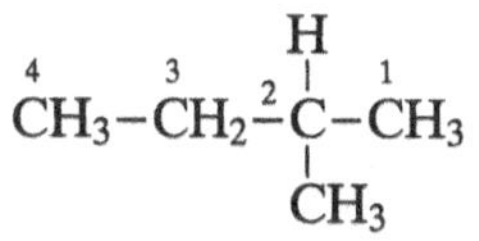

$$\underset{\underset{\text{CH}_3}{|}}{\overset{\overset{\text{H}}{|}}{\overset{4}{\text{CH}}_3-\overset{3}{\text{CH}}_2\overset{2}{-}\text{C}-\overset{1}{\text{CH}}_3}}$$

The IUPAC names are 2-methylpropane (compound **a**), 3-methylpentane (compound **b**), and 2-methylbutane (compound **c**).

If two or more substituents are present, we must do several things: (1) number the chain of the parent to give the substituted carbon the lowest number; (2) give each substituent a locant (position) number; and (3) if there is more than one of the same substituent, provide a prefix before the substituent to indicate how many substituents of this type are present. These prefixes have Latin or Greek roots: di (2), tri (3), tetra (4), penta (5), hexa (6), and so on.

Q Name the following compounds:

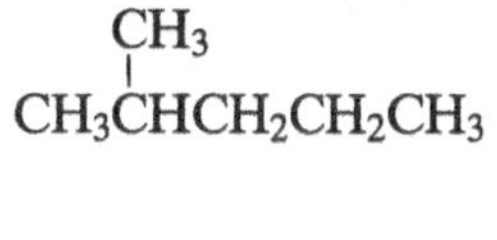

$$\underset{\underset{}{}}{\overset{\overset{\text{CH}_3}{|}}{\text{CH}_3\text{CHCH}_2\text{CH}_2\text{CH}_3}} \qquad \underset{\underset{\text{CH}_3}{|}}{\overset{\overset{\text{CH}_3}{|}}{\text{CH}_3\text{CCH}_2\text{CH}_2\text{CH}_3}}$$

(a) 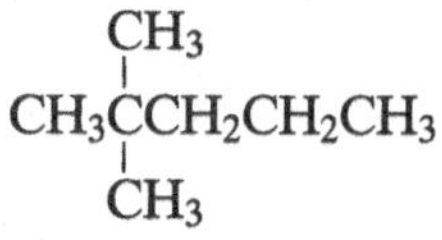 (b)

A Compound **a** has five carbons as its longest chain and one substituent—the methyl group. We must number the chain to give the substituent the lowest possible number, and therefore this compound is 2-methylpentane. Compound **b** is also a pentane, but it has two substituents, both of which are methyl groups. The parent chain must also be numbered from the left, and each methyl group must be assigned the number 2. The name therefore is 2,2-dimethylpentane. The names 2-dimethylpentane and 2,2-methylpentane are *not* correct.

Q Provide a line formula for 3-ethyl-2,2-dimethylpentane. Before you set pen to paper to draw the formula, note that the substituents are given in alphabetical order without regard to any prefix. The format for the hyphens and commas is also important.

A

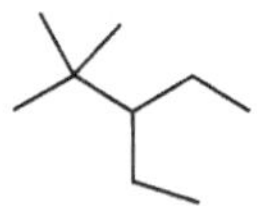

Other Substituents. Other types of groups can also function as substituents, including

- The halogens—F, fluoro; Cl, chloro; Br, bromo; and I, iodo
- The OH, hydroxy group
- The CN, cyano group
- The NO$_2$, nitro group
- The NH$_2$, amino group

The compound CH$_3$Cl is chloromethane (also called *methyl chloride*); the compound CHCl$_3$ is trichloromethane (commonly called *chloroform*); CH$_2$Cl$_2$ is dichloromethane (usually called *methylene chloride*); and CH$_3$CHBrCH$_3$ is 2-bromopropane.

Q Name the following compound:

$$\text{F-CH}_2\text{CH}_2\overset{\overset{\displaystyle\text{CH}_3}{|}}{\underset{\underset{\displaystyle\text{CH}_3}{|}}{\text{C}}}\text{CH}_3$$

A The longest chain contains four carbons, and the parent name is therefore butane. There are three substituents—one fluoro group and two methyl groups. Each of these must be given a number corresponding to the carbon to which they are attached. There are two ways to number the carbon base. If we number from the right-hand carbon, the name would be 4-fluoro-2,2-dimethylbutane. If we number from the left side, the name would be 1-fluoro-3,3-dimethylbutane.

Statement. According to the IUPAC rules, the parent should be numbered beginning at the end nearer the first branch or substituent point. This rule results in a name that has the lowest possible number for a substituent.

Thus, the second name is the correct name. ■

Q Give a condensed formula for the compound 1,1,1-trichloro-3-nitropropane.

A Cl$_3$CCH$_2$CH$_2$NO$_2$. ■

Branched Hydrocarbon Substituents. Branched hydrocarbons can also function as substituents, and there are two ways to name these groups. We will call these two methods the *common* method and the *IUPAC* method, even though there is some overlap between these two methods. Table 2.3 provides the common name for straight-chain and branched substituents containing three and four carbons.

The isopropyl and propyl groups differ in the point of attachment in the group. The middle carbon of the isopropyl group is attached to the parent hydrocarbon. The compound shown below is 4-propyloctane.

$$\text{CH}_3\text{CH}_2\text{CH}_2\overset{\overset{\displaystyle\text{CH}_2\text{CH}_2\text{CH}_3}{|}}{\text{CH}}\text{CH}_2\text{CH}_2\text{CH}_2\text{CH}_3$$

Q Give the name of the compound:

$$\text{CH}_3\text{CH}_2\text{CH}_2\overset{\overset{\displaystyle\text{CH}_3\text{CHCH}_3}{|}}{\text{CH}}\text{CH}_2\text{CH}_2\text{CH}_3$$

A 4-isopropylheptane. ■

TABLE 2.3. Three and Four-Carbon Substituents

Formula	Name
Three-Carbon Substituents	
CH$_3$CH$_2$CH$_2$—	propyl
CH$_3$CHCH$_3$ (attachment at middle C)	isopropyl
Four-Carbon Substituents	
CH$_3$CH$_2$CH$_2$CH$_2$—	butyl
CH$_3$CH$_2$CHCH$_3$ (attachment at 2nd C)	*secondary* butyl or *sec*-butyl
CH$_3$CHCH$_2$— with CH$_3$ branch	isobutyl
CH$_3$CCH$_3$ with CH$_3$ branch	*tertiary* butyl or *tert*-butyl or *t*-butyl

The difference between the butyl groups is more subtle but can be understood by carefully looking at the difference between the groups and by also understanding the meaning of the words *primary*, *secondary*, and *tertiary*. Let's start with the words.

When only one carbon is attached to the carbon that will substitute for hydrogen on the parent, the group is a *primary* group. If two carbons are attached to that substitutive carbon, the group is *secondary*, and if three carbons are attached, it is *tertiary*. For example, each of the groups shown below is primary:

$$CH_3CH_2CH_2- \quad CH_3CH_2CH_2CH_2- \quad \begin{array}{c} CH_3CHCH_2- \\ | \\ CH_3 \end{array}$$

Each of the groups below has a secondary carbon of attachment:

$$\begin{array}{c} | \\ CH_3CHCH_3 \end{array} \quad \begin{array}{c} | \\ CH_3CH_2CHCH_3 \end{array}$$

Each of the groups below has a tertiary carbon at the attachment site:

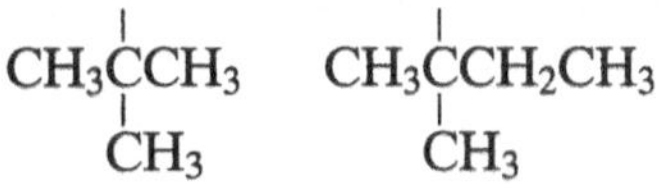

$$\begin{array}{c} | \\ CH_3CCH_3 \\ | \\ CH_3 \end{array} \quad \begin{array}{c} | \\ CH_3CCH_2CH_3 \\ | \\ CH_3 \end{array}$$

Now let's think about all the possible ways that four carbons can be arranged.

Q Write a structure for all of the possible four-carbon alkanes.

A

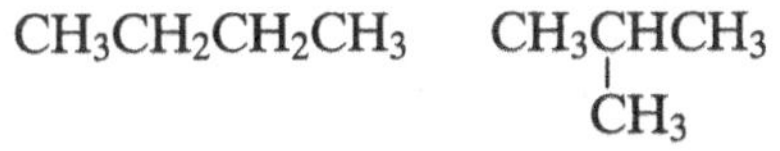

$$CH_3CH_2CH_2CH_3 \quad \begin{array}{c} CH_3CHCH_3 \\ | \\ CH_3 \end{array}$$

Let's take a look at butane to see which hydrogens can be replaced by a substituent. There are only two carbons at which substitution can occur:

$$\underset{\underset{\text{here}}{\uparrow} \quad \underset{\text{or here}}{\uparrow}}{CH_3CH_2CH_2CH_3}$$

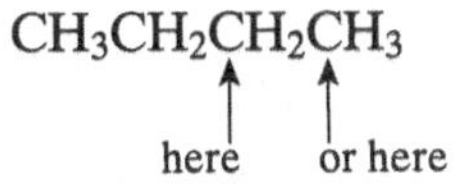

Q Consider the other possible four-carbon alkane and determine where substitution can occur.

A

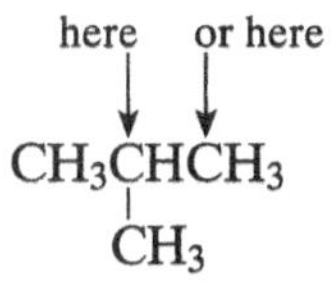

$$\overset{\text{here} \quad \text{or here}}{\underset{\downarrow \quad \downarrow}{CH_3CHCH_3}} \\ \begin{array}{c} | \\ CH_3 \end{array}$$

Remember that substitution at any of the three terminal atoms gives an equivalent group. Be sure that you see that the four substitution sites on the two alkanes correspond to the four butyl groups. ∎

Q Write the formula and give the name of the butyl group derived from 2-methylpropane that has a terminal substitutive carbon.

A

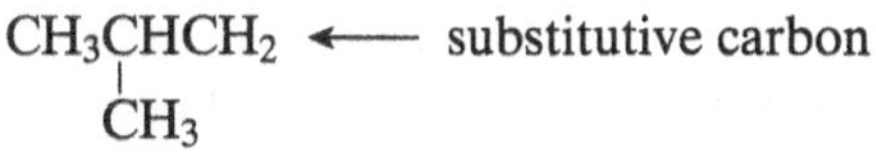

$$\begin{array}{c} CH_3CHCH_2 \\ | \\ CH_3 \end{array} \longleftarrow \text{substitutive carbon}$$

Note that the carbon that attaches (the substituting or *substitutive carbon*) is a primary carbon, just like the substitutive carbon in the butyl group. This branched group is named the *isobutyl group*. ∎

Alkanyl Names. In the *IUPAC method*, the group is named using the same rules that apply to any hydrocarbon, except that an additional number is required to indicate the substitutive carbon. The name of the group is also modified to indicate that it is a substituent and follows the generic name *alkanyl*. This method sounds a bit complicated but looks a lot simpler when applied to some substituents. For example, the isopropyl group

$$\begin{array}{c} | \\ CH_3CHCH_3 \end{array}$$

has the name propan-2-yl, illustrating that the number of the attachment, or substitutive, carbon is given immediately before the -yl. The *tert*-butyl group

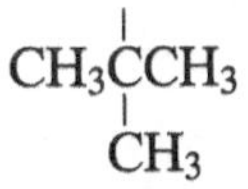

$$\begin{array}{c} | \\ CH_3CCH_3 \\ | \\ CH_3 \end{array}$$

can be named 2-methylpropan-2-yl.

Q Give the alkanyl name for the group

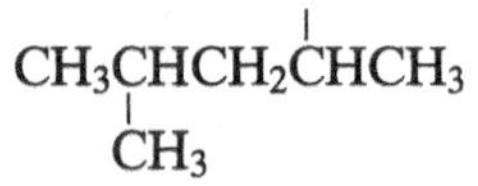

$$\text{CH}_3\text{CHCH}_2\overset{|}{\text{CHCH}_3}$$
$$\overset{|}{\text{CH}_3}$$

A 4-methylpentan-2-yl. Note that the carbons are numbered to give the attachment carbon precedence over the methyl group. ■

When the attachment carbon is the terminal carbon of the substituent, such as $\text{CH}_3\text{CH}_2\text{CH}_2$–, the preferred name is the common name, in this case propyl. For most other groups the alkanyl name is preferred. For the $(\text{CH}_3)_3\text{C}$– substituent, the common name, *tert*-butyl, is generally used.

Cycloalkanes. Alkanes also may be cyclic compounds. These cyclic hydrocarbons are known as *cycloalkanes*, and their names are based on the number of carbons in the ring. Thus, a ring with three carbons is cyclopropane, one with four carbons is cyclobutane, one with five carbons is cyclopentane, and so on. Like normal alkanes these compounds may have substituents. Consider the following examples:

methylcyclopropane   1-bromo-2-ethylcyclohexane

1-methyl-1-propylcyclopentane

Q Draw structures for two different cyclic molecules that contain a total of four carbons.

A

Note that cyclic compounds have two fewer hydrogens than do their *acyclic* (the prefix *a* means *not*, just as the word *apolitical* means *not political*) analogs. ■

Q Give the molecular formula for cyclobutane and its acyclic analog.

A The molecular formula for cyclobutane is C_4H_8; for butane, C_4H_{10}. The cyclic analog has two fewer hydrogens than the acyclic parent. ■

Q Name the following compound:

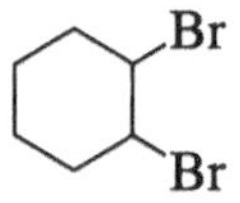

A In this compound the longest carbon chain contains six carbons; thus, the compound is a cyclohexane. There are two substituents: the two bromo groups. The cyclohexane must be numbered to assign one substituent to carbon 1. The compound is 1,2-dibromocyclohexane. ■

Q Give line formulas for *t*-butylcyclopentane and cyclopropylcyclobutane.

A

Note that in the second name a cyclic alkane is used as a substituent. ■

Alkenes

Alkenes contain at least one carbon–carbon double bond. The simplest alkene, $\text{H}_2\text{C}=\text{CH}_2$, has the common name *ethylene*. The IUPAC name of ethylene is *ethene*, which is derived from the name of the analogous alkane, *ethane*, by replacing the *-ane* with *-ene*. For more complicated alkenes, the position of the double bond in the longest carbon chain must be indicated by a number. The most recent IUPAC recommendation is to place the number before the *-ene* ending. For example, $\text{CH}_3\text{CH}=\text{CHCH}_3$ is but-2-ene because the first carbon in the double bond is the second carbon in the chain.

The formula CH₃CH₂CH₂CH=CH₂ denotes pent-1-ene, rather than pent-4-ene, because the position of the double bond is given the lowest number possible. It is also important to become familiar with the slightly older format that places the number of the double bond before the parent name. Thus, CH₃CH=CHCH₃ can also be named 2-butene.

 Name the following compounds

 3-chlorobut-1-ene (3-chloro-1-butene) and 1,4-dichlorobut-2-ene (1,4-dichloro-2-butene). ■

 What is wrong with the name 1,4-dimethylbut-2-ene?

 This name produces the formula

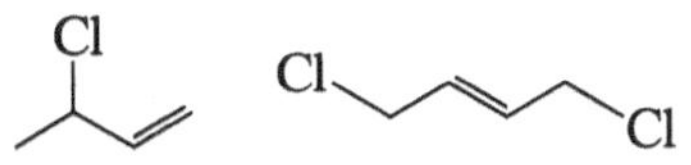

CH₃CH₂CH=CHCH₂CH₃

Of course, this is 3-hexene and not a butene, because the parent alkene has six carbons. ■

Cycloalkenes have a carbon–carbon double bond within the ring, and these two carbon atoms are assigned as number one and number two in naming compounds:

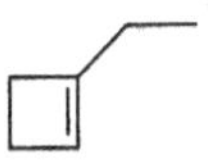

1-ethylcyclobutene 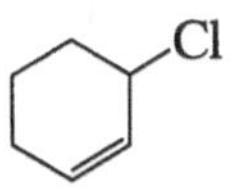   3-chlorocyclohexene

 What is wrong with the name 5-aminocyclopentene?

 This name conjures up the formula

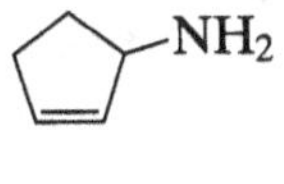

which should be named 3-aminocyclopentene because the substituent must be given the lowest possible number and the numbering of the carbons in the ring must start at one of the double-bonded carbons as shown below.

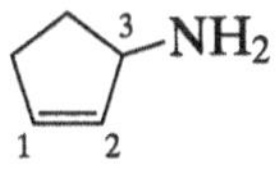

■

 The cyano group is probably not well known to you. It always has the carbon attached to the parent hydrocarbon. Write a Lewis structure for cyanomethane, more commonly called *acetonitrile*.

H
|
H–C–C≡N:
|
H

■

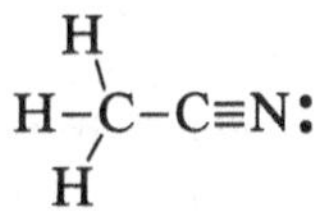

Alkene Geometric Isomers. The rigid nature of the double bond results in isomers for some alkenes. We will discuss isomerism in more detail later, but for now consider the following two structures and names for the isomers of but-2-ene (2-butene):

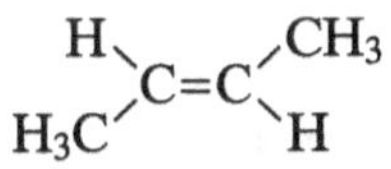

trans-2-butene 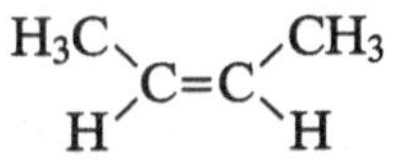   cis-2-butene

In the first structure the longest chain of carbon atoms includes two carbons that are on opposite sides of the double bond, and this arrangement of the hydrogens is referred to as the *trans* isomer. In the second structure the corresponding hydrogens are on the same side of the double bond, and the compound is the *cis* isomer. (To help you remember these prefixes, think of the meaning of *transatlantic*, meaning across the Atlantic, or the many other words that have this Latin prefix.)

 Draw the line formula for *cis*-hex-3-ene.

■

When each carbon of the double bond contains only one hydrogen the *cis/trans* nomenclature provides an unambiguous designation of the relative orientation of the groups. However, in a compound such as

$$\underset{H}{\overset{H_3C}{>}}C=C\underset{Br}{\overset{Cl}{<}}$$

it is not clear whether this isomer should be designated cis or trans. The *E, Z* system of nomenclature was designed to deal with these situations by assigning a priority to each group on each carbon of the double bond. The details of the priority system will be discussed in your organic chemistry course, but for now it will be helpful to know that higher priorities are assigned to atoms with higher atomic numbers. Thus, in the example above, Br has a higher priority than Cl on the right-hand carbon, and CH_3 has a higher priority than H on the left-hand carbon. If the two high-priority groups are on the same side of the molecule, the isomer is designated as the *Z* isomer (the word *zusammen* is German, meaning *together*); if the two high priority groups are on opposite sides of the double bond, the isomer is designated the *E* isomer (*entgegen*, meaning *opposite*). Thus, the isomer above is the *E* isomer of 1-bromo-1-chloro-1-propene. The following compound is (*Z*)-3-bromo-2-pentene:

Alkenes as Substituents. Alkenes can also be named as substituents. The IUPAC system approves the use of the *alkenyl* system, analogous to the alkanyl system, or the names *vinyl* and *allyl* for the groups $CH_2=CH-$ and $CH_2=CH-CH_2-$, respectively. In the alkenyl system, the chain is numbered to give the carbon attached to the parent the lowest possible number, and that number is placed before the *–yl*. The number of the double bond is placed before the *–en*. Thus, the allyl group $CH_2=CH-CH_2-$ is named prop-2-en-1-yl, and the vinyl group $H_2C=CH-$ is simply ethenyl (there are only two carbons and numbers are therefore unnecessary to specify the position of either the carbon of attachment or the double bond in the vinyl group). The following group is prop-1-en-2-yl:

We will have occasion to use these groups with aromatic compounds and compounds with functional groups.

Benzene is a very special type of hydrocarbon that appears to be a cyclic alkene. This compound does not undergo reactions typical of alkenes, however, and for that reason and others the structural formula that shows three discrete double bonds is often an inadequate description of the bonding in this compound.

We have already discussed the bonding in benzene. Benzene and similar compounds are discussed later as aromatic compounds or arenes.

Alkynes

The IUPAC rules for naming *alkynes*, those hydrocarbons that contain carbon–carbon triple bonds, are identical to the alkene rules except that the *-ane* ending of the parent alkane is replaced by *-yne* to indicate the presence of the triple bond. Propyne is $CH_3C\equiv CH$; the simplest alkyne, ethyne ($HC\equiv CH$), also has the common name acetylene. Two alkynes with the molecular formula C_4H_6 exist:

$$H-C\equiv C-CH_2CH_3 \qquad H_3C-C\equiv C-CH_3$$
$$\text{1-butyne} \qquad\qquad \text{2-butyne}$$

Because the triple bond is at the end of the chain, but-1-yne (1-butyne) is called a *terminal* alkyne. The compound but-2-yne (2-butyne) is an *internal* alkyne.

Q Name the following alkyne:

$$H-C\equiv C-CCl_3$$

A The triple bond must be given the lowest possible number. Therefore, the carbon chain is numbered from left to right, and the compound is 3,3,3-trichloroprop-1-yne or 3,3,3-trichloro-1-propyne. Notice that the 1 refers to the carbon where the triple bond begins. The name 3,3,3-trichloropropyne is also acceptable because it leads to an unambiguous structure. Because the carbon bonded to three halogen substituents can form only one other bond, it is obvious that the triple bond occurs between the other two carbon members of the chain. Both names are therefore correct. ∎

Q Name the following compounds:

CH₃CH₂−C≡C−CH₂CH₃   Br−C≡C−CH₂ĊHCH₂CH₃
                                              |
                                             CH₃

A The first compound is hex-3-yne (or 3-hexyne), an example of a symmetric alkyne. The second compound is 1-bromo-4-methylhex-1-yne. ∎

Aromatic Compounds or Arenes

Compounds that contain the benzene ring (discussed in Chapter 1) are called *aromatic* compounds (the compounds often have a distinct odor). Three different ways of drawing the benzene ring are shown below:

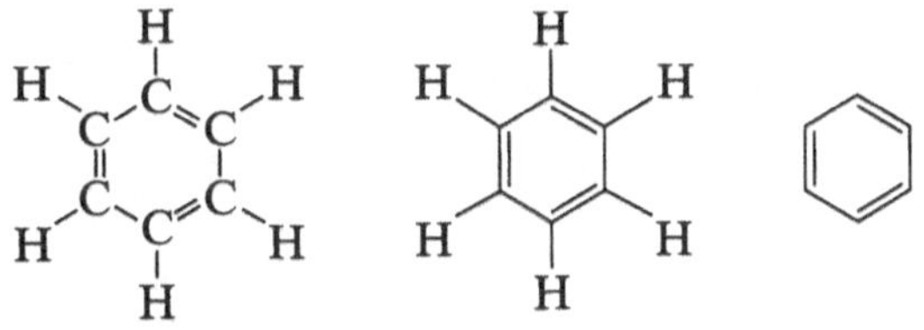

Note that the presence of the carbons is understood in the middle structure and the presence of both carbon and hydrogen atoms is understood in the line structure. An alternative structural description of the bonding in the compound uses a ring to indicate that the double bonds interact with each other to produce additional stability in benzene relative to a cyclic compound with three discrete double bonds. This representation is rarely used today, but it is important to recognize its meaning when one reads older literature or textbooks in which the notation was used. Two alternate representations of benzene are

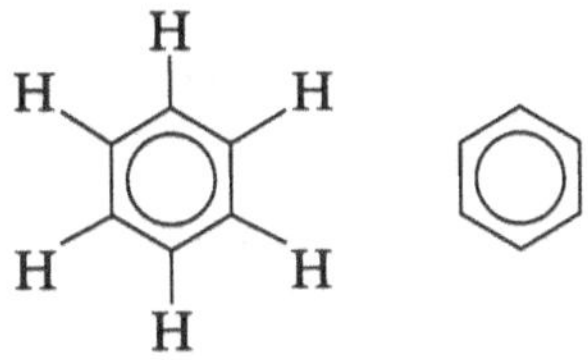

When one hydrogen on benzene is replaced by another group, the remaining C₆H₅− is referred to as a *phenyl* group, as, for example, in phenylacetylene.

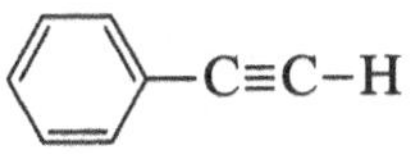

Moreover, the phenyl group's contribution to the molecule's structure is called the *aryl* portion of the compound.

Q Write a line structural formula for *trans*-1-phenyl-2-pentene.

A ∎

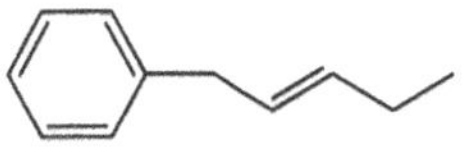

Q Name this same compound as a substituted benzene.

A *trans*-pent-2-en-1-ylbenzene or simply *trans*-pent-2-enylbenzene (the 1 is implied). ∎

Substituted benzenes (benzenes in which substituents such as halogens or an alkyl group replace one or more of the hydrogen atoms) are named by the usual IUPAC rules. For example

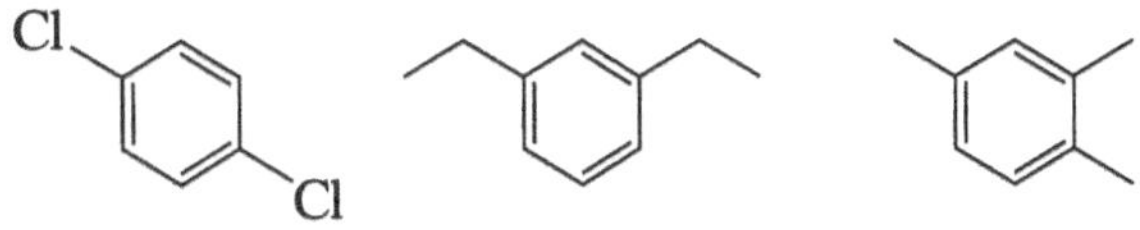

1,4-dichlorobenzene   1,3-diethylbenzene   1,2,4-trimethylbenzene

Q Give a structure for 1,3,5-trinitrobenzene.

A

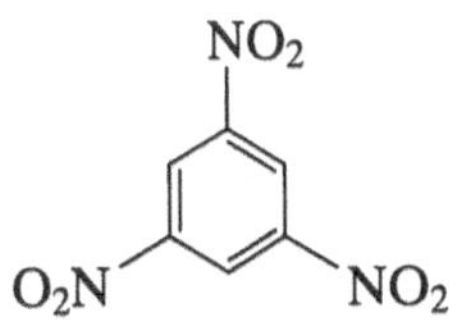

Note that this structural formula clearly indicates that it is the nitrogen of the nitro group that is attached to the carbon of the benzene ring. ∎

Substituted aromatics can also be named by the *common system*, in which adjacent substituents are designated by the prefix *ortho* (*o-*), substituents two atoms removed are designated by *meta-* (*m-*), and substituents directly opposite are designated by *para-* (*p-*), as illustrated below.

Q Provide a common name for the moth repellant ("mothballs" may contain either the compound below or another aromatic compound, naphthalene):

A *para*-dichlorobenzene. ∎

Q An important compound closely related to benzene is methylbenzene, which is more commonly called *toluene*.

Name the following compounds using both the IUPAC and common methods to indicate the positions of substituents.

A The first compound is 4-chlorotoluene or *para*-chlorotoluene. The second compound may be named either 2-bromotoluene or *ortho*-bromotoluene. ∎

Q A toluene derivative made famous by its explosive power is 2,4,6-trinitrotoluene (TNT). This compound is actually not as sensitive to shock as most people believe and was originally used as a yellow dye. Like many nitro compounds, it is fairly toxic and on prolonged exposure turns the skin a yellow-orange color. Write a formula for TNT.

A ∎

Q Provide names for each of the following:

A 1-bromo-2-ethylbenzene or *ortho*-bromoethylbenzene, 1-bromo-4-iodobenzene or *para*-bromoiodobenzene, 1,3,5-triethylbenzene, 1,2-dichloro-4-fluorobenzene, and 1,2,4,5-tetramethylbenzene. ∎

Many compounds contain the benzene ring along with one of the functional groups that we will discuss in the next section. For example, phenol is an aromatic alcohol, benzaldehyde is an aromatic aldehyde, and benzoic acid is an aromatic carboxylic acid:

phenol   benzaldehyde   benzoic acid

There are also compounds that do not contain a simple benzene ring but do possess aromatic characteristics. These compounds include the following *polycyclic* compounds, all of which can be obtained from coal tar. Some of these polycyclic aromatics are carcinogenic (cause cancer). In fact, benzopyrene was the first carcinogen to be identified. All of them can also bear substituents, and all of them consist of fused benzene rings.

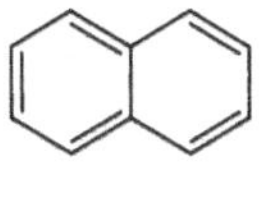

naphthalene 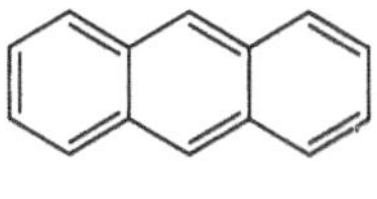   anthracene 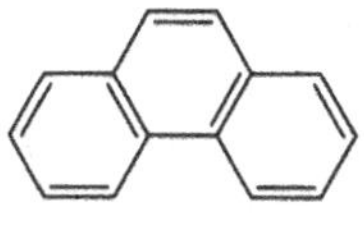   phenanthrene

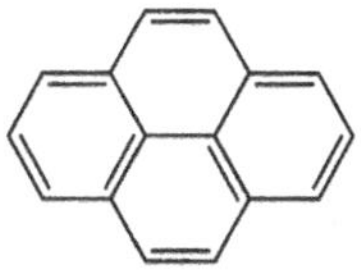

pyrene 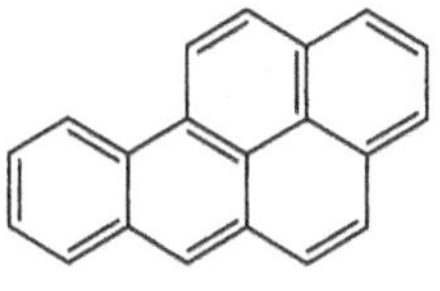   benzopyrene

FUNCTIONAL GROUPS

Certain substituents or groups, called *functional groups*, give organic molecules characteristic chemical behaviors that are very different from those of hydrocarbons. For example, when an –OH group replaces hydrogen in a structure, some reagents that would not react with the parent hydrocarbon will react with the –OH group. Note that without the OH group in the center of the compound below (an alcohol) it would be a simple hydrocarbon:

$$\text{CH}_3\text{CH}_2\text{CH}_2\overset{\overset{\displaystyle\text{OH}}{|}}{\text{CH}}\text{CH}_2\text{CH}_2\text{CH}_3$$

Alcohols

Alcohols contain hydroxyl (OH) functional groups bonded to tetrahedral carbon atoms. The general formula for alcohols is ROH, in which R represents a hydrocarbon group.

Q What is R in the compound $CH_3CH_2CH_2OH$?

A The propyl group.

Q What is R in the compound

$$\text{CH}_3\text{CH}_2\overset{\overset{\displaystyle\text{OH}}{|}}{\text{CH}}\text{CH}_3 ?$$

A The group

$$\text{CH}_3\text{CH}_2\overset{|}{\text{CH}}\text{CH}_3$$

is known in the common system as the *sec*-butyl group. Thus, a common name for the compound is *sec*-butyl alcohol.

In the IUPAC system, alcohols are named by first determining the longest chain of carbon atoms *that contains the OH group*. The name is produced from the name of the parent hydrocarbon by replacing the *-e* ending by *-ol*. The hydrocarbon chain is numbered to give the OH group the lowest number. For the simplest alcohol CH_3OH the name is based on the parent alkane, methane, and the name is therefore methanol.

Q Give the name of the compound CH_3CH_2OH that is present in alcoholic beverages and is often present in the gasoline that we use in our cars.

A ethanol.

For more complex alcohols the position of the OH group on the carbon chain is indicated by a number in front of the *-ol* (in some texts the number is placed before the parent name). Thus, the saturated three-carbon alcohol in which the hydroxyl group is on the second carbon atom is propan-2-ol or 2-propanol.

$$\text{CH}_3\overset{\overset{\displaystyle\text{OH}}{|}}{\text{CH}}\text{CH}_3 \quad \text{2-propanol}$$

If the position of the alcohol functional group could be indicated by two different location numbers, then the name that assigns the lower number to the hydroxyl position is used. For example, in the structure shown below the proper name for the compound would be 6-methylheptan-3-ol and not 2-methylheptan-5-ol.

$$\text{CH}_3\overset{\overset{\displaystyle\text{CH}_3}{|}}{\text{CH}}\text{CH}_2\text{CH}_2\overset{\overset{\displaystyle\text{OH}}{|}}{\text{CH}}\text{CH}_2\text{CH}_3$$

Q Give a line formula for the compound 2-butanol.

A

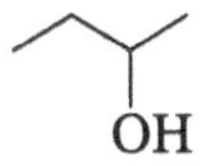

(line formula showing 2-butanol with OH)

Q Provide a name for the compound

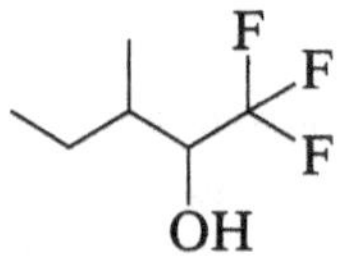

A 1,1,1-trifluoro-3-methylpentan-2-ol or 1,1,1-trifluoro-3-methyl-2-pentanol.

Q Reason by analogy with the rules for naming alkenes to name the compound $CH_2=CH–CH_2OH$. [*Hint*: The hydroxyl group is given priority (the lowest number) in the numbering of the hydrocarbon chain.]

A prop-2-en-1-ol.

Q Draw a line formula for 2,2,4,4-tetramethylpentan-3-ol.

A

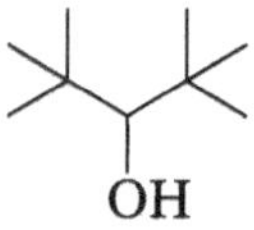

Q Draw a line formula for 3-methyl-1-butanol.

A

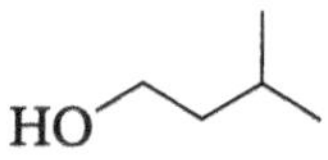

Cyclic alcohols are named using the same rules. Therefore, the compound represented by both structures below would be named 2-bromocyclohexanol.

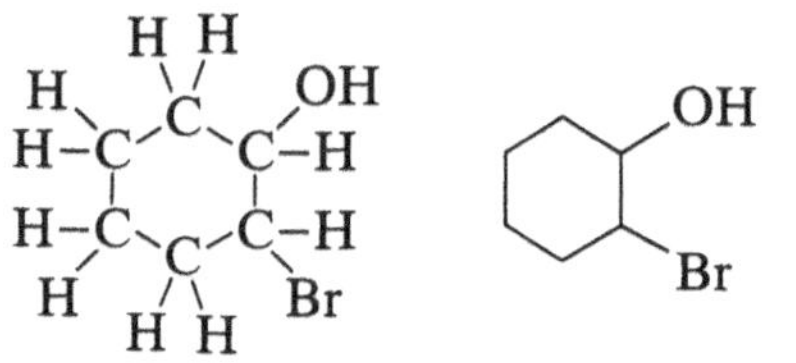

Q The compound

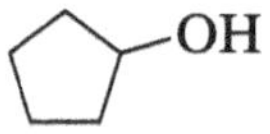

is cyclopentanol. Name the compound

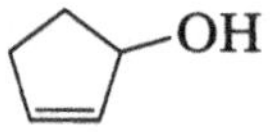

A cyclopent-2-enol, or cyclopent-2-en-1-ol.

Many compounds containing functional groups are also named by the common system. In this system the carbon skeleton is named as a group, rather than a hydrocarbon parent. In the case of alcohols the group name is followed by the word *alcohol*. For example, CH_3CH_2OH is named *ethanol* in the IUPAC system and in the common system, *ethyl alcohol*. The names of the three- and four-carbon common groups were summarized earlier (see Table 2.3).

Q Give both the IUPAC and common names for the compound

A 2-methylpropan-2-ol (2-methyl-2-propanol) and *tert*-butyl alcohol.

Q Provide the IUPAC name for isopropyl alcohol.

A 2-propanol (propan-2-ol).

Q Provide the common name for cyclopentanol.

A cyclopentyl alcohol.

Q Name the following compounds:

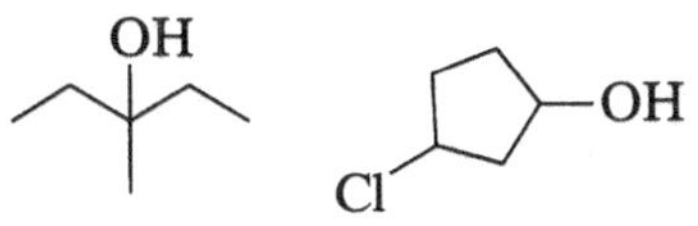

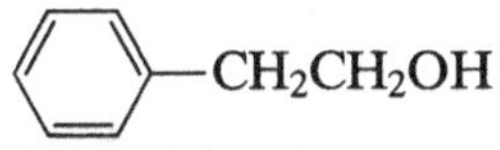

A 3-methylpentan-3-ol (3-methyl-3-pentanol), 3-chlorocyclopentanol, 2-phenylethan-1-ol (2-phenyl-1-ethanol), but-3-yn-1-ol (3-butyn-1-ol).

Q If a compound has two identical functional groups the prefix *di-* is used in the name to indicate this circumstance. Name the following compounds:

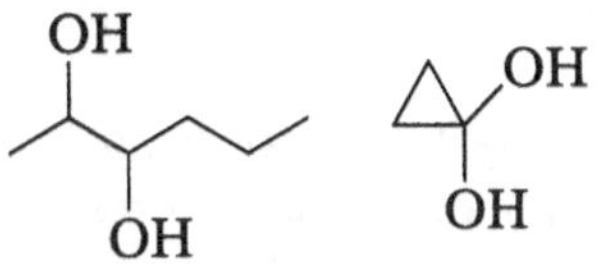

A hexane-2,3-diol and cyclopropane-1,1-diol.

Phenols

Phenols are compounds that have hydroxyl groups attached to aromatic rings. These compounds are more acidic than alcohols. Phenols occur widely in nature and are used in many industrial preparations of important compounds. The parent compound (phenol) may be used as a disinfectant. Various substituted phenols may be used as flavoring agents, and some naturally occurring phenols are the compounds in poison ivy that produce allergic reactions.

The nomenclature of substituted phenols is based on the assignment of position 1 to the carbon that bears the hydroxyl group. The structures below are examples. The assignment of position one is implied in the names.

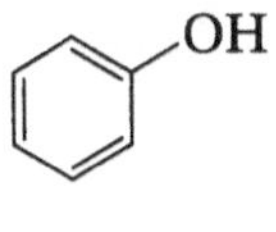

phenol

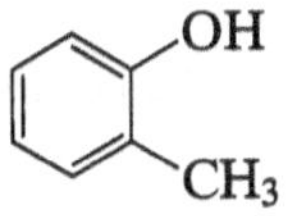

2-methylphenol
o-methylphenol

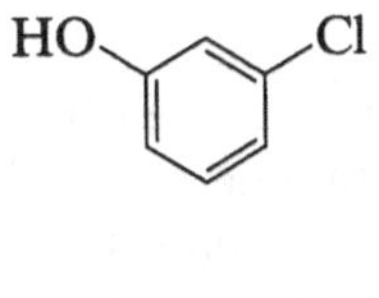

3-chlorophenol
m-chlorophenol

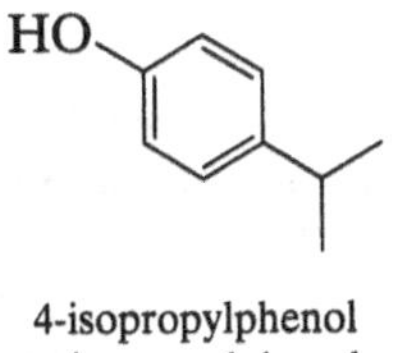

4-isopropylphenol
p-isopropylphenol

Q What are the names of the following compounds?

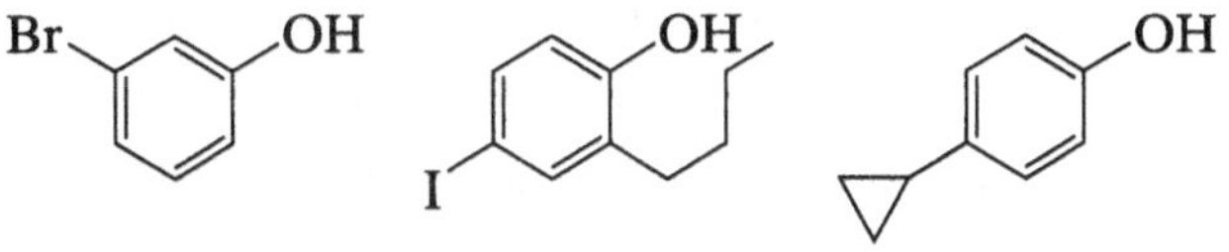

A The first structure represents 3-bromophenol or *m*-bromophenol. The second compound is 2-butyl-4-iodophenol, and the third compound is 4-cyclopropylphenol, or it may be called *p*-cyclopropylphenol.

Q Use the alkanyl system to name the compound

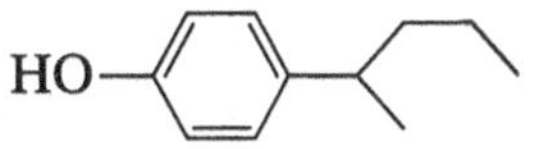

A 4-(pentan-2-yl)phenol.

Ethers

Ethers are compounds in which two alkyl groups are attached to a central oxygen atom. The general structure is

$$R-O-R'$$

In symmetric ethers R and R' are the same, and in asymmetric ethers R and R' are different groups.

Q Use the *tert*-butyl group as R and the isopropyl group as R' to formulate an asymmetric ether.

A The formula is $(CH_3)_3C-O-CH(CH_3)_2$, where we have used a slightly different notation for the two groups, but by now you should recognize the groups as

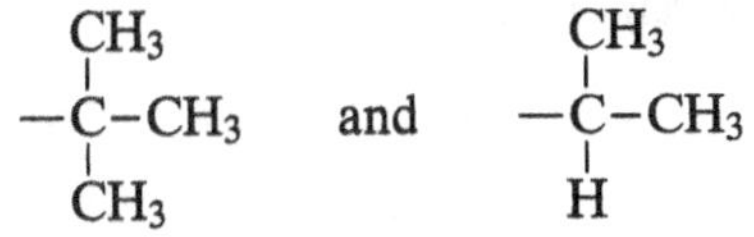

$$\begin{array}{c} CH_3 \\ -C-CH_3 \\ CH_3 \end{array} \quad \text{and} \quad \begin{array}{c} CH_3 \\ -C-CH_3 \\ H \end{array}$$

The IUPAC rules allow two methods for naming ethers. If the compound is a simple ether, then it is named by identifying the two organic groups followed by the word *ether*. The common solvent frequently called "ether" is actually diethyl ether.

CH₃CH₂–O–CH₂CH₃   H–C(CH₃)(CH₃)–O–CH₃

diethyl ether   methyl isopropyl ether

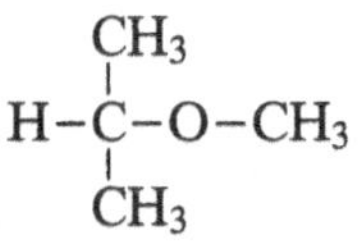

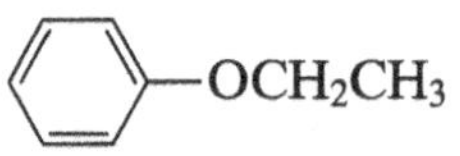

⌬—OCH₂CH₃

ethyl phenyl ether

If more than one ether functionality is present or if other functional groups are present, then the ether is named as an alkoxy substituent by replacing the *-yl* ending of a group with *-oxy*. For example, the CH₃O– group is the methoxy group.

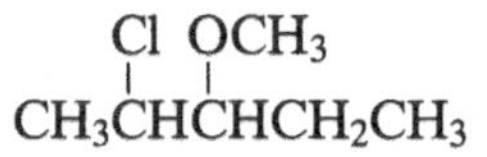

Cl OCH₃
CH₃CHCHCH₂CH₃   [cyclopentene]—OCH₂CH₃

2-chloro-3-methoxypentane 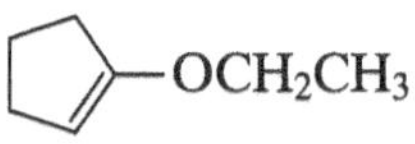   1-ethoxycyclopentene

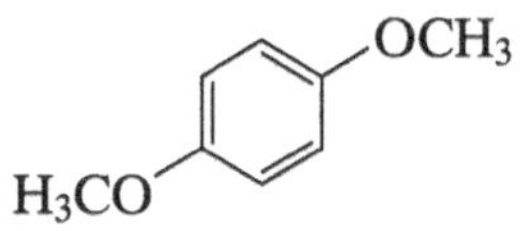

H₃CO—⌬—OCH₃

1,4-dimethoxybenzene

Q Provide appropriate names for the following compounds:

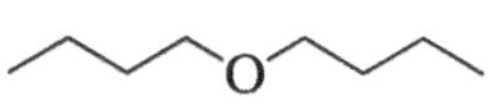

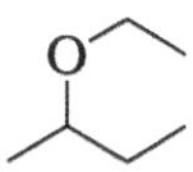

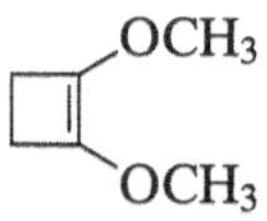

[structures]

A The names are dibutyl ether, 2-ethoxybutane (or *sec*-butyl ethyl ether), and 1,2-dimethoxycyclobutene. ∎

Ketones and Aldehydes

Ketones have the following generic formula:

R–C(=O)–R

where R ≠ H.

The group of atoms between the two R groups, C=O, is the carbonyl (car-bo-*neel*) group, and it is present in many of the most important functional groups.

Q The R groups in a ketone do not have to be the same. In the simplest ketone both R groups are methyl. Do you remember this compound? Draw a structure for it and name it.

A

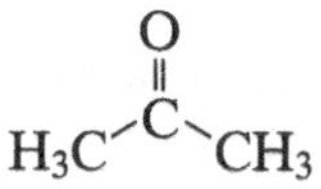

H₃C–C(=O)–CH₃

is propanone or acetone (trivial name). ∎

In the IUPAC system ketones are named by locating the longest chain of carbon atoms that contains the carbonyl group. The *-e* of the parent carbon alkane is replaced by *-one* to designate a ketone. The ketone in the question above is named propanone (as you surely remember!). The compound

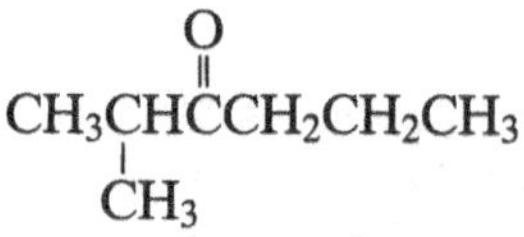

CH₃CHCCH₂CH₂CH₃
         ‖
     CH₃ O

contains six carbons in the longest chain and one substituent, a methyl group. As before, the chain must be numbered so that the carbonyl carbon receives the lowest possible number. Therefore, we number from the left side of the given structure, and the number is placed before the *-one* ending. In some texts the number is placed before the parent name. The compound is named 2-methylhexan-3-one or 2-methyl-3-hexanone.

Q Name the compound

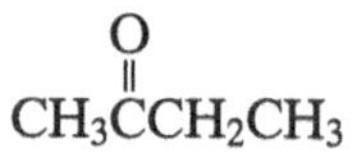

CH₃CCH₂CH₃
   ‖
   O

A Because the carbonyl group can be only at the second position the name butanone, rather than butan-2-one, is sufficient. This is a common solvent known as methyl ethyl ketone in the common system. It is frequently abbreviated MEK. ∎

Q Give a line formula for 4-hydroxybutan-2-one (4-hydroxy-2-butanone).

A

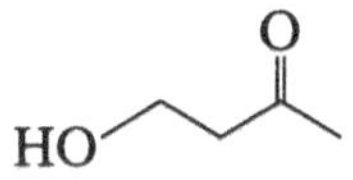

[structure: HO–CH2–CH2–C(=O)–CH3]

Q Name the compound

[structure: (CH3)2CH–CH2–C(=O)–CH2–CH(CH3)2]

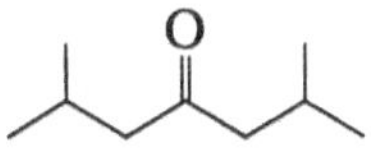

A 2,6-dimethylheptan-4-one (2,6-dimethyl-4-heptanone).

In the common system the two groups attached to the carbonyl are named and are then followed by the word *ketone*. In this system, butan-2-one would be named methyl ethyl ketone (see above). Similarly, the compound shown below is cyclopropyl methyl ketone.

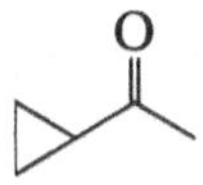

[structure: cyclopropyl–C(=O)–CH3]

Q Name the compound

[structure: CH2=CH–C(=O)–CH(CH3)2]

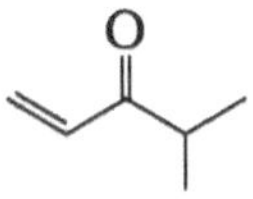

A 4-methylpent-1-en-3-one or isopropyl vinyl ketone (the vinyl group is –CH=CH2).

Q Give a formula for *tert*-butyl isobutyl ketone.

A

(CH3)3CCCH2CH(CH3)2
       ‖
      O

Q Give the IUPAC name for the ketone above.

A 2,2,5-trimethylhexan-3-one (2,2,5-trimethyl-3-hexanone).

Q Name the following:

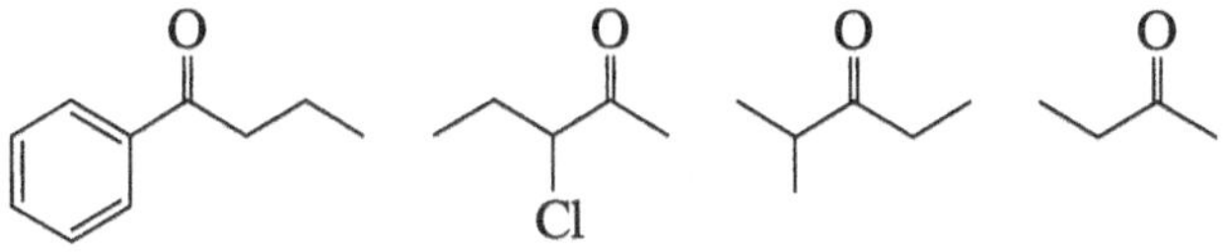

[four structures: PhC(=O)CH2CH2CH3; CH3CH2C(=O)CH(Cl)CH3; (CH3)2CHC(=O)CH2CH3; CH3C(=O)CH2CH3]

A 1-phenylbutan-1-one or phenyl propyl ketone, 3-chloropentan-2-one, 2-methylpentan-3-one or ethyl isopropyl ketone, butanone or methyl ethyl ketone (frequently referred to as MEK).

Aldehydes have a carbonyl group with the following general structure:

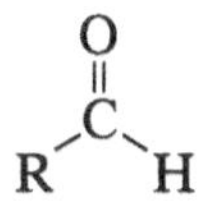

[structure: R–C(=O)–H]

They are generally more reactive than ketones because the carbonyl group is slightly more polar in an aldehyde than it is in a similar ketone. Also, because the hydrogen atom is smaller than an R group, reagents more readily attack the carbonyl carbon atom.

Q Draw a line formula for an aldehyde in which R = CH3CH2.

A

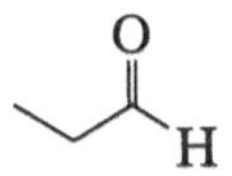

[structure: CH3CH2–C(=O)–H]

Note that for aldehydes, R can be H. The compound H2CO is the simplest aldehyde. For ketones, R cannot be H (if one of the R groups on a ketone were a hydrogen, the compound would be an aldehyde).

As was the case in naming ketones, the carbonyl group is given priority in a numbering scheme. In the

IUPAC system aldehydes are named by locating the longest chain of carbon atoms that contains the carbonyl group. The -e of the base carbon alkane is replaced by –al to designate an aldehyde. Thus, a straight-chain aldehyde with five carbons would be pentanal. Four low-molecular-weight aldehydes with their IUPAC names and common names are

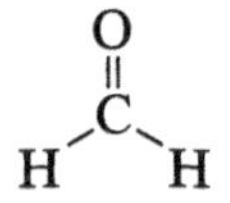

methanal
formaldehyde

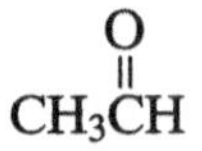

ethanal
acetaldehyde

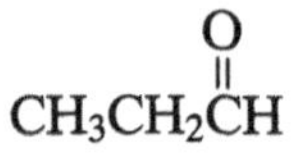

propanal
propionaldehyde

butanal
butyraldehyde

As we will see in the next section, the common names of aldehydes are derived from the name of the corresponding carboxylic acid.

Cyclic compounds that are aldehydes are named by stating the name of the ring system followed by the term *carbaldehyde*. When the carbon of the aldehyde's carbonyl group is attached to a benzene ring, the compound is named *benzaldehyde*.

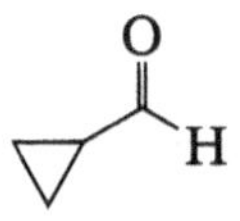

cyclopropanecarbaldehyde

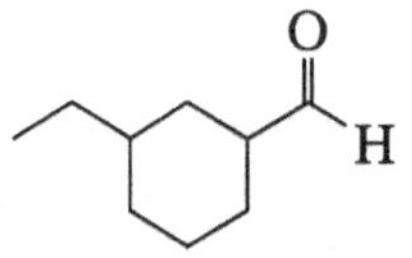

3-ethylcyclohexanecarbaldehyde

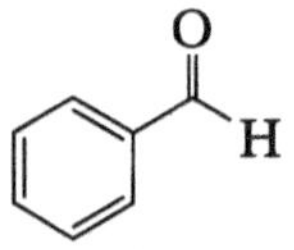

benzaldehyde

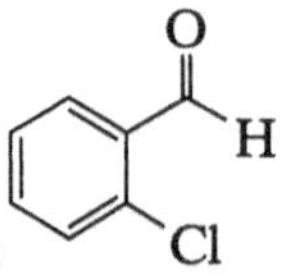

2-chlorobenzaldehyde
o-chlorobenzaldehyde

Q What are the names of the carbonyl compounds shown below?

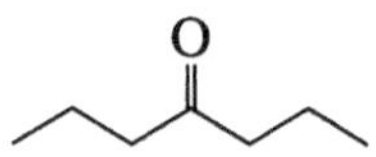

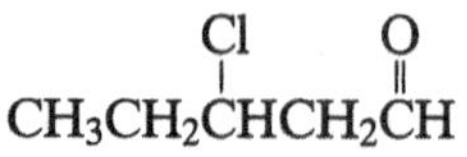

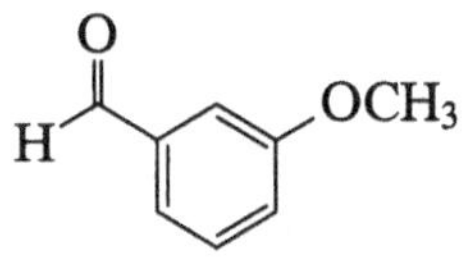

$CH_3CH_2CHCH_2CH$ (with Cl on middle CH and =O on terminal CH)

A heptan-4-one, 3-chloropentanal, and 3-methoxybenzaldehyde, which may also be named *m*-methoxybenzaldehyde. ∎

Q Give the name of the following compound:

$CH_3CCH_2CCH_3$ (with two =O groups)

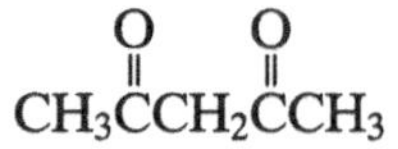

A As the presence of two identical functional groups in a molecule may be indicated by the prefix di-, this compound would be named 2,4-pentanedione (pentane-2,4-dione). ∎

Q Propose a name for the following compound:

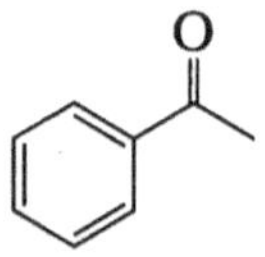

A With the carbonyl group's higher priority, the benzene ring would be considered a substituent. Thus, the compound could be named 1-phenylethanone or methyl phenyl ketone. It is more commonly called acetophenone. ∎

Carboxylic Acids

Carboxylic acids have the following generic formula:

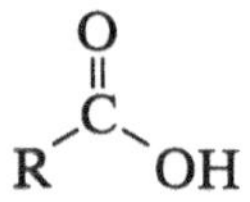

$$R-C(=O)-OH$$

In these compounds the carbonyl carbon is attached to a hydroxyl group. Because of the electronegativity of both oxygens, carboxylic acids release the OH hydrogen as a hydrogen ion or proton to molecules that have a lone pair of electrons. Compounds that function in this way are called *acids*.

Q Write a line structural formula for the acid obtained when R is the phenyl group.

A

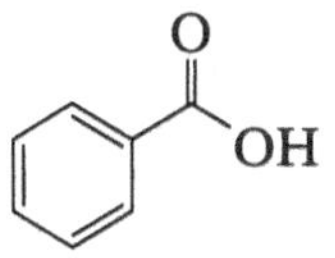

In the IUPAC system the name of the compound is based on the longest carbon chain that includes the carbonyl carbon. The *c* of the parent alkane is then replaced by *-oic acid*. For example, the compound

$$CH_3\underset{\underset{OH}{|}}{C}HCH_2\overset{\overset{O}{\|}}{C}OH$$

is 3-hydroxybutanoic acid.

Q Give the IUPAC name for the compound with an isobutyl group as R.

A 3-methylbutanoic acid.

Acids can also be named in the common system on the basis of names related to the origin of the acid. For example, methanoic acid can be obtained by distilling the bodies of ants, and because the Latin name for ant is *formica*, this acid is known as *formic acid*. The IUPAC and common names for the simple carboxylic acids are given in Table 2.4.

These common names can be used in a variety of acid derivatives and have already been used as the basis for the common names of the aldehydes.

If the carboxylic acid functionality is a substituent on a ring, then the compound is named using the parent ring name and adding carboxylic acid to it. If the carboxylic acid functionality is a substituent on benzene, then the compound is named as a benzoic acid.

TABLE 2.4. IUPAC and Common Names for Carboxylic Acids

IUPAC Name	Common Name	Formula
methanoic acid	formic acid	HCO_2H
ethanoic acid	acetic acid	CH_3CO_2H
propanoic acid	propionic acid	$CH_3CH_2CO_2H$
butanoic acid	butyric acid	$CH_3CH_2CH_2CO_2H$
pentanoic acid	valeric acid	$CH_3CH_2CH_2CH_2CO_2H$

Q Provide names for the structures below.

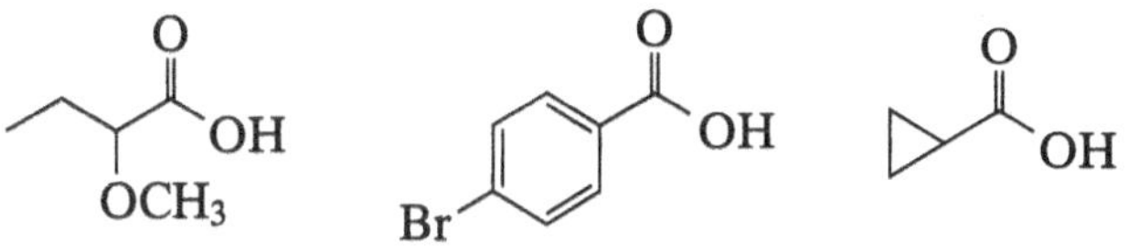

A 2-methoxybutanoic acid, 4-bromobenzoic acid (or *p*-bromobenzoic acid), and cyclopropane carboxylic acid.

Q Write line structures for 2-nitropropanoic acid, 3-cyanobutanoic acid, and 2-aminoethanoic acid. Note that these three compounds contain the NO_2 (nitro), CN (cyano), and the NH_2 (amino) groups. The CN and NH_2 groups are also the nitrile and amino functional groups, but here it is necessary to treat them as substituents.

A

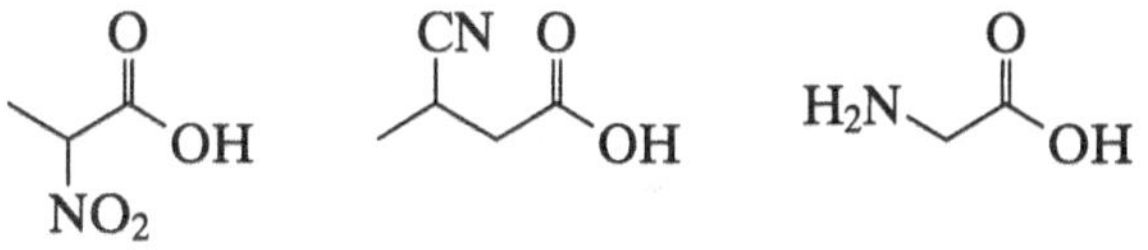

Acid Derivatives

Esters. Esters are derivatives of carboxylic acids in which the H of the acid has been replaced by either an alkyl group or an aryl group. They are often characterized by their sweet or fruity odor.

Q Provide the generic structure for an ester.

A

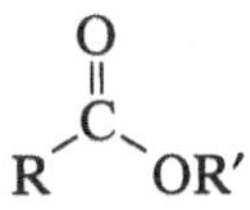

$$R-\overset{\overset{O}{\|}}{C}-OR'$$

Q Note that in the generic structural formula above the R' group in effect has replaced the hydrogen that would be present on the acid RCO₂H. Write a structure for the compound obtained when the hydrogen of propanoic acid is replaced by an ethyl group. (This process can be achieved in the laboratory by reacting an alcohol containing the R′ group with an acid containing the R group.)

A

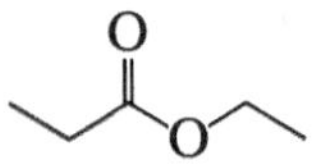

Esters are named by first giving the name of the group that replaced the H followed by the name of the acid with the *-oic acid* ending replaced by *-ate*. For the compound whose structure you just wrote, the group that replaced the H is the ethyl group and the name of the acid is propanoic acid in the IUPAC system or propionic acid in the common system. Thus, the name of this compound is either ethyl propanoate or ethyl propionate, depending on the name you choose for the acid.

The compound below may be named as either ethyl ethanoate or ethyl acetate:

$$\text{CH}_3\overset{\overset{\text{O}}{\|}}{\text{C}}\text{OCH}_2\text{CH}_3$$

Q Use the common name for the parent acid to name the compound CH₃CO₂CH₃.

A The common name for the parent acid is acetic acid, and therefore this compound is methyl acetate.

Other examples include

$$\text{CH}_3\text{CH}_2\text{CH}_2\text{CH}_2\overset{\overset{\text{O}}{\|}}{\text{C}}\text{OCH}_3$$

methyl pentanoate
methyl valerate

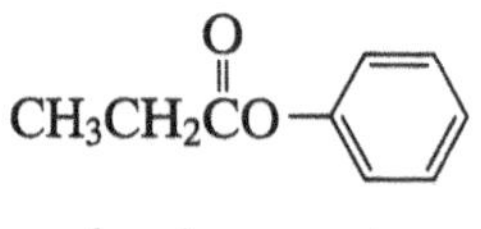

$$\text{CH}_3\text{CH}_2\overset{\overset{\text{O}}{\|}}{\text{C}}\text{O}\!\!-\!\!\bigcirc$$

phenyl propanoate
phenyl propionate

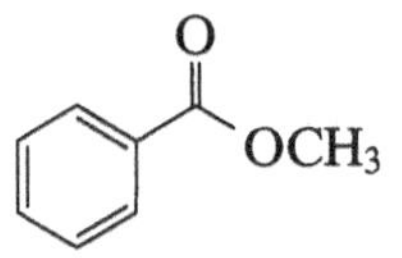

methyl benzoate

Q Give a structural formula for the ester derived from propanoic acid by replacing the acidic proton with an isopropyl group and name it.

A

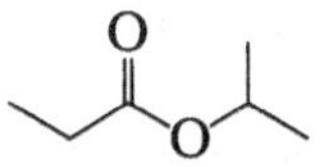

The compound is isopropyl propanoate.

Q Write line formulas for *tert*-butyl acetate, propyl 3-chlorobutanoate, and 7-bromooctyl ethanoate.

A

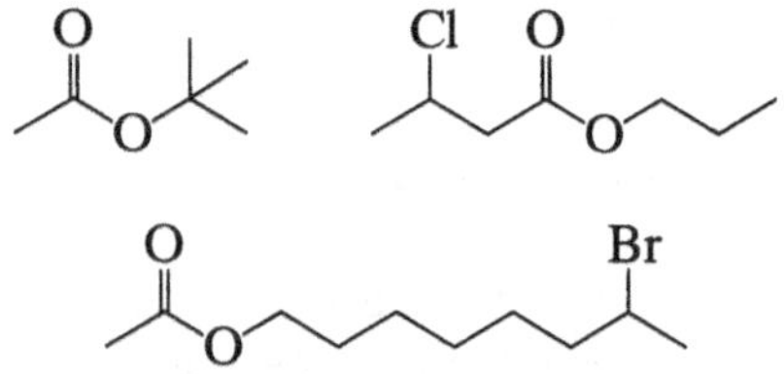

Acid Anhydrides, Acid Halides, Amides, and Nitriles. Acid anhydrides are formed by condensing two molecules of an acid with the removal of water. The formula below shows the OH of one acid molecule combining with the H of another to form a compound that contains two acyl groups attached to an oxygen.

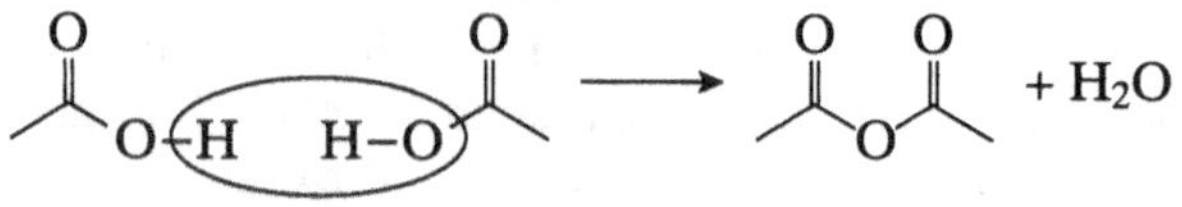

Here two molecules of acetic acid condense to form acetic anhydride. The names of acid anhydrides are based on the names of the parent acid with the word *acid* replaced by the word *anhydride*.

Q What is the name of the CH₃CO– group?

A The acetyl group, named after the acid CH₃CO₂H, acetic acid. This is an acyl group, which has the generic formula where R can be either an alkyl or an aryl group. In IUPAC nomenclature, the same group is named ethanoyl after ethanoic acid.

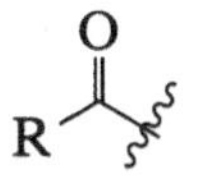

Q Give the name and structure of the anhydride formed from benzoic acid.

A benzoic anhydride:

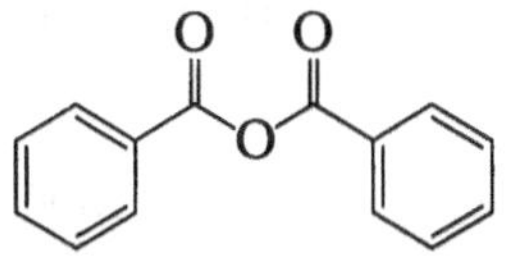

Acyl halides, like CH_3COCl, contain a halide attached to the acyl group and are named as acyl halides. For example, CH_3COCl is acetyl chloride or ethanoyl chloride.

Amides contain an NH_2 group (or NHR or NR_2 group; see below under the heading Amines) attached to the acyl group and are named by replacing the *-yl* part of the name of the acyl group with *amide*. For example, CH_3CONH_2 is acetamide.

Q Name the amide $C_6H_5CONH_2$.

A benzamide:

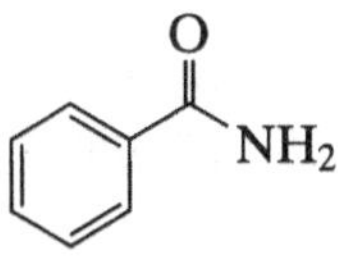

Nitriles are not analogous to carboxylic acids but are frequently obtained from acids. They have the generic formula RCN. If $R = CH_3$, the compound is named acetonitrile, a common solvent: $CH_3-C\equiv N$.

Amines

Organic derivatives of ammonia (NH_3) are known as *amines*. These compounds are structurally derived from ammonia just as ethers and alcohols are structurally derived from water. Amines are commonly found in plants and animals in three general forms, with increasing numbers of organic groups attached to the central nitrogen atom.

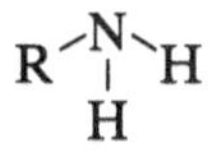

Primary amine 1° amine
Secondary amine 2° amine 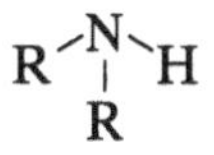
Tertiary amine 3° amine 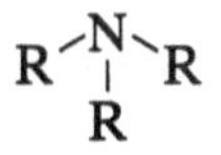

Q Write a formula for a secondary amine that has two $R = CH_3$ groups.

A

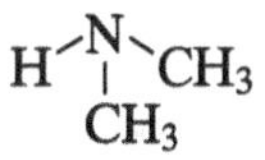

Amines may be named by adding the suffix *amine* to the alkyl or aryl group(s) name(s), or they may be named by replacing the *-e* ending of the parent compound's name with *-amine*. For example, the amine with one methyl group attached to the nitrogen could be called methylamine or methanamine. If more than one group is attached to the nitrogen, then they must be clearly identified in the name. For example, if two ethyl groups are bonded to the nitrogen atom, then the name would be diethylamine, and if three phenyl groups are bonded to the same nitrogen, then the appropriate name would be triphenylamine.

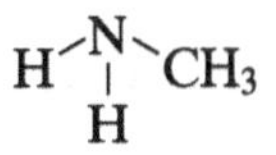

methylamine
methane amine

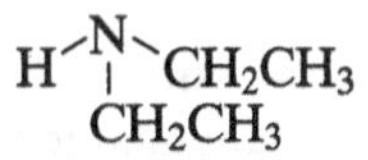

diethylamine

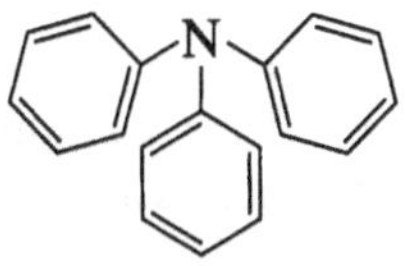

triphenylamine

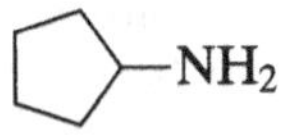

cyclopentylamine
cyclopentanamine

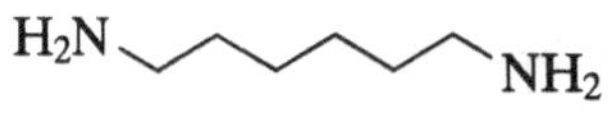

1,6-hexanediamine
1,6-diaminohexane

Secondary and tertiary amines that are asymmetric are named as *N*-substituted primary amines in which the largest group is chosen to determine the parent name of the compound. The prefix *N*- indicates that a substituent is bonded directly to the nitrogen atom.

Q Provide formulas for *N*-methylpropylamine and *N*, *N*-diethylcyclohexylamine.

NOMENCLATURE

A

CH₃NCH₂CH₂CH₃
  |
  H

(cyclohexyl-N(ethyl)(ethyl))

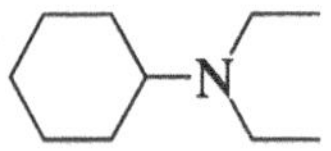

Amines that have other functional groups are usually named with the –NH₂ group considered to be an *amino* substituent. For example, the compounds below would be 4-amino-2-pentanone and 3-aminopropanoic acid, respectively.

$$\underset{\text{NH}_2}{\text{CH}_3\text{CHCH}_2\text{CCH}_3} \quad \overset{\text{O}}{\underset{\|}{\text{C}}}\quad \text{HOCCH}_2\text{CH}_2\text{NH}_2$$

Q The α-amino acids are an important class of compounds because they form the backbone of peptides and proteins. The word *alpha* or symbol α designates the carbon next to the carbonyl of the acid functionality. An α-amino acid has the generic formula

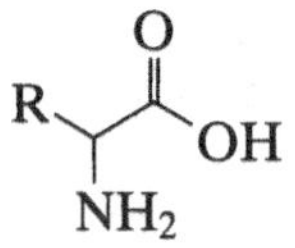

$$\underset{\text{NH}_2}{\text{R}}\!\!\diagdown\!\!\overset{\text{O}}{\underset{}{\text{C}}}\!\!\diagup\text{OH}$$

Give the IUPAC name of the compound derived from this formula where R = CH₃.

A This amino acid is 2-aminopropanoic acid. It is generally known by the name alanine.

Benzene with an attached amino group is usually called aniline, although it could also be named aminobenzene or benzamine. Aniline may have substituents on the aromatic ring or on the nitrogen or both.

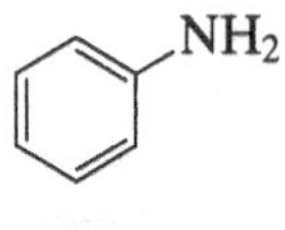

aniline 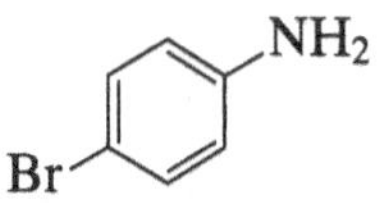   4-bromoaniline / *p*-bromoaniline

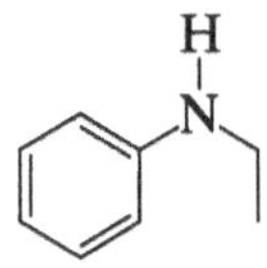

N-ethylaniline 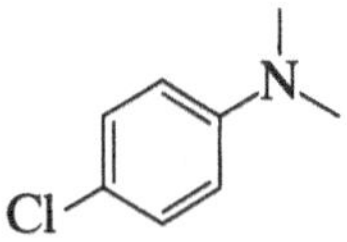   4-chloro-*N,N*-dimethylaniline / *p*-chloro-*N,N*-dimethylaniline

CUMULATIVE NOMENCLATURE PROBLEMS

Q Name the following compounds:

(a) CH₃CH₂CH₃   (b) CH₃CH₂CH₂CHCH₃
                              |
                             CH₃

(c)  \\==/   (d) CH₃C≡CH

(e) CH₃CCH₃   (f) CH₃CH
       ‖   ‖
       O   O

(g) HCOH   (h) CH₃OCH₂CH₃
     ‖
     O

(i) CH₃CH₂NH₂

A (a) propane; (b) 2-methylpentane; (c) *cis*-2-butene; (d) propyne; (e) propanone (acetone); (f) ethanal or acetaldehyde; (g) methanoic acid or formic acid; (h) ethyl methyl ether; (i) ethylamine or ethanamine.

Q Write structural formulas for each of the following compounds:

(a) propene   (b) propyne   (c) propanone
(d) propanal   (e) propanoic acid (f) 2-propanol

A

(a) 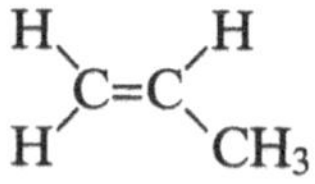 H\C=C/H with H and CH₃   (b) H–C≡C–CH₃

(c) CH₃CCH₃   (d) HCCH₂CH₃
         ‖   ‖
         O   O

(e) HOCCH₂CH₃   (f) CH₃CHCH₃
         ‖ |
         O   OH

Q Provide names for the following compounds:

(a) [structure: phenyl-CH2-CH2-CH(OH)-CH3]

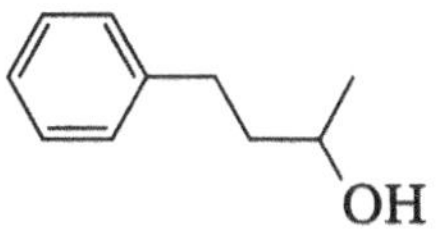

(b) [structure: CH3-CH(NH2)-CH2-CH2-CH2-COOH]

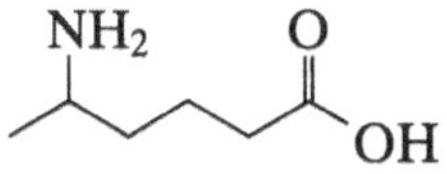

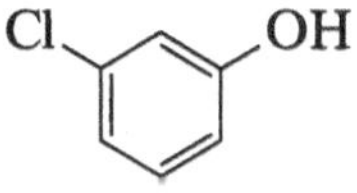

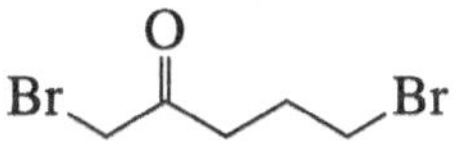

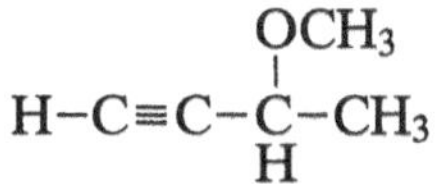

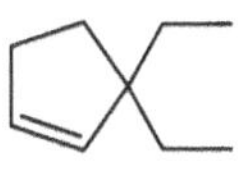

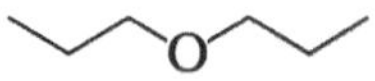

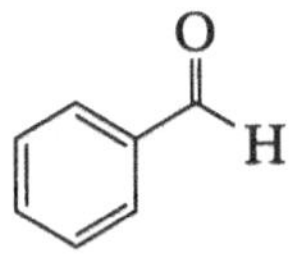

(c) [structure: 3-chlorophenol]

(d) [structure: Br-CH2-C(=O)-CH2-CH2-CH2-Br]

(e) H−C≡C−C(OCH3)(H)−CH3

(f) [structure: cyclopentene with two ethyl groups at C3]

(g) [structure: CH3CH2-O-CH2CH2CH3 — dipropyl ether]

(h) [structure: benzaldehyde]

(i) [structure: trans-2-pentene]

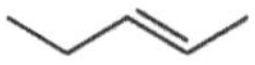

A

(a) 4-phenylbutan-2-ol; (b) 5-aminohexanoic acid; (c) 3-chlorophenol or *m*-chlorophenol; (d) 1,5-dibromopentan-2-one; (e) 3-methoxybut-1-yne; (f) 3,3-diethylcyclopentene; (g) dipropyl ether; (h) benzaldehyde; (i) *trans*-pent-2-ene or (*E*)-2-pentene. ∎

3

BONDING

The way in which atoms are held together within a molecule is still not completely understood despite almost a century of effort devoted to an understanding of the mechanics of atom attachment. The models used to describe these atomic interactions vary in complexity from the oldest model, the Lewis model, to the molecular orbital model, which requires extensive calculations but is capable of describing the bonding and structure of even fairly large molecules. The extent to which each of these models produces predictions that agree with experimental molecular parameters, such as bond lengths, bond angles, and dipole moments, determines the usefulness of the model. Although the Lewis model is not capable of the precise predictions possible with the molecular orbital model, it is extremely easy to use. Consequently, it has become a tool with which all students of chemistry must be familiar. The proper use of this model permits chemists to rationalize the ways in which organic molecules react. Consequently, we will spend most of our time on it and its wave mechanical relative, the valence bond model. It is essential that you work diligently on all of the exercises provided in order to become proficient in drawing and manipulating Lewis structures.

THE LEWIS MODEL

This model was proposed by the American chemist Gilbert Newton Lewis (1875–1946) in the early twentieth century and is based on the fact that stable ions generally have the electron configuration of an inert gas. For example, the chloride ion has one more electron than the atom, which gives it the electron configuration of argon. Lewis reasoned that this generalization might be true not only for ionic compounds like NaCl or MgO but also for covalent compounds such as water or organic molecules.

Q Remembering that Lewis formulas contain only valence electrons and that the electrons are represented by dots and when possible the electrons are shown as pairs, which of the following is a Lewis formula for the chloride ion?

$$:\!\ddot{Cl}\!\cdot \quad :\!\ddot{Cl} \quad :\!\ddot{Cl}\!:$$

A The first is the neutral chlorine atom with seven valence electrons. The last one with eight valence electrons is correct for the chloride ion. ∎

Molecules do not contain ions, however, and Lewis proposed that the electrons in a molecule arrange themselves in order to provide an inert-gas configuration for each atom in the molecule. Of course, this is a very simple model that treats the electrons as particles that can move about to create these configurations. Although it has many disadvantages that stem from its simplicity, many of its basic assumptions are used in the most sophisticated current models. In order to create the "electronic cement" that holds the atoms together, Lewis suggested that there must be at least a pair of

---

The Bridge to Organic Chemistry: Concepts and Nomenclature
By Claude H. Yoder, Phyllis A. Leber, and Marcus W. Thomsen
Copyright © 2010 John Wiley & Sons, Inc.

electrons positioned directly between each set of attached atoms.

Lewis also realized that only the electrons in the *valence shell* of each atom are held loosely enough to be involved in the bonding between atoms. The electrons below the valence shell are very tightly bound to the nucleus. Lewis structures therefore show only the valence electrons in the molecule. Thus, in order to write Lewis structures, we must know the number of valence electrons for each atom in a molecule.

Q Write the electron configuration for a carbon atom.

A The electron configuration of an atom is a description of which orbitals contain electrons. The orbital with the lowest energy is the $1s$ orbital, which is the only orbital in the first quantum level (for which the quantum number $n = 1$). In the second quantum level (where $n = 2$) there is a total of four orbitals—the $2s$, and three $2p$ orbitals. These five orbitals can hold a total of 10 electrons (two per orbital). Carbon has six electrons, and we place these electrons into the lowest-energy orbitals:

$$C\ 1s^2 2s^2 2p^2$$

The highest-energy electrons for carbon are actually in two separate p orbitals (following Hund's first rule), which is not obvious from the way we have written the electron configuration.

Q How many valence electrons are there for carbon?

A The electrons in the lowest-energy orbital, the $1s$ orbital, are very strongly attracted to the nucleus and are therefore not involved in bonding. Only the electrons in the highest quantum level are the valence electrons. Therefore, for carbon only the electrons in the $2s$ and $2p$ orbitals are valence electrons. Carbon has four valence electrons. ∎

In general this number of valence electrons can be obtained from the periodic chart. For example, the elements in group 14, referred to as "group IV" on some periodic charts, have four electrons in the valence shell, no matter whether the element is carbon, silicon, germanium, tin, or lead. Group 15 elements have five valence electrons, and so on. Of course, hydrogen has only one valence electron.

Q How many valence electrons do boron and oxygen have?

A Boron is in group 13 and has three valence electrons. Oxygen is in group 16 and has six valence electrons. ∎

The electron configuration of an inert gas, except helium (group 18), will have eight electrons in its valence shell. Consequently, the Lewis model is often known as the *octet model*, or we say that a molecule must obey the *octet rule*. There are exceptions to this rule that we will discuss later, but we must recognize that hydrogen cannot possibly have eight electrons in its valence shell. Its valence shell, the $1s$ orbital, can hold only two electrons; too much energy would be required to use the second quantum level ($n = 2$) to obtain room for more than these two electrons. Thus, helium has a pair of electrons in its valence shell.

Lewis structures are easily written by using the following procedure:

1. Write the atoms of the molecule in the positions in which they appear in the molecule. The hydrogen atoms in methane (CH_4), are situated at the corners of a tetrahedron (the angles between the C–H bonds are $109°$), and the carbon sits at the center of the tetrahedron. When the structural formula of methane is written, it is not always given the three-dimensional perspective of the tetrahedron, but is usually represented with the flat look shown below:

$$\begin{array}{c} H \\ | \\ H-C-H \\ | \\ H \end{array}$$

The important point is that this formula tells us that the Lewis structure that we write must show at least a pair of electrons positioned directly between each hydrogen atom and the central carbon. The formula must also give each hydrogen two electrons in its valence shell, while the carbon must "see" eight electrons.

Q Do not confuse structural formulas with Lewis structures, which in this case happen to be exactly the same. Draw a structural formula for methanol.

A The formula

$$\text{H}\cdots\underset{\underset{\text{H}}{|}}{\overset{\overset{\text{H}}{|}}{\text{C}}}\cdots\text{O}\diagdown\text{H}$$

is intended only to show us the atom connectivities. ■

2. Count up the total number of valence electrons for the methane molecule. Methane has four valence electrons for carbon and one for each of the four hydrogen atoms, to give a total of eight valence electrons for the whole molecule.

3. Arrange these eight electrons to satisfy the Lewis criteria. Because of the importance of the Lewis "electronic cement" criterion (i.e., that we have at least one pair of electrons between each set of attached atoms), we place two electrons between each set of attached atoms. For methane, this means that we draw a line, representing two electrons, between the carbon and each hydrogen atom. Each of these pairs of electrons is called a *single bond*.

4. Rearrange the electrons if necessary, to give each atom an octet of electrons. In counting the electrons for this second criterion of Lewis, we include all of the electrons in the bonds to each atom. These electrons are, after all, shared by the atoms. In methane, the eight electrons in the four C–H bonds give the carbon a total of eight electrons, thereby satisfying the octet rule. The two electrons in each C–H bond also give each hydrogen atom a duet of electrons. Thus, the Lewis structure for methane is

$$\text{H}-\underset{\underset{\text{H}}{|}}{\overset{\overset{\text{H}}{|}}{\text{C}}}-\text{H}$$

Q Determine the Lewis structure for methanol.

A The structural formula is given in the previous answer. The total number of valence electrons is 4 (C) + 4 × 1 (4 H) + 6 (O) = 14. Placing two electrons between each set of connected atoms produces the electron dot structure

$$\text{H}-\underset{\underset{\text{H}}{|}}{\overset{\overset{\text{H}}{|}}{\text{C}}}-\text{O}-\text{H}$$

This structure does not have a total of 14 electrons, however; it contains only 10 electrons. Thus, four electrons must be placed somewhere in the structure. They cannot be placed on a hydrogen, nor on the carbon, because each hydrogen already has two electrons in its molecular valence shell and carbon has eight. There is only one place to put the four electrons—on the oxygen. These four electrons are placed in two pairs on the oxygen. ■

It may seem strange to place the electrons in pairs around an atom; they do, after all, repel one another. This pairing of electrons has a quantum-mechanical justification: electrons with spin quantum numbers that are different (remember that according to quantum mechanics each electron is characterized by four quantum numbers—n, l, m, and s—and that in a given atom no two electrons can have exactly the same set of four quantum numbers) can approach one another *more closely* than electrons that have the same value for the spin quantum number. The complete Lewis structure is

$$\text{H}-\underset{\underset{\text{H}}{|}}{\overset{\overset{\text{H}}{|}}{\text{C}}}-\overset{..}{\underset{..}{\text{O}}}-\text{H}$$

The two pairs of electrons on the oxygen are not involved in bonding and are therefore referred to as *nonbonded* electrons or as *lone pair* electrons.

Q Write an electron dot formula for the common ketone CH$_3$COCH$_3$ (our old friend propanone or acetone).

A The structural formula for this ketone contains two methyl groups attached to the carbonyl carbon. There are 24 valence electrons [3 × 4(C) + 6 × 1 (H) + 1 × 6 (O) = 24] in acetone, and of these, 12 are used up in the C–H bonds (each methyl group contains three C–H bonds). The other 12 must be distributed in the bonds between the methyl groups and the central carbon, in the bond between carbon and oxygen, and as nonbonded (lone pair) electrons on the oxygen. After we distribute the other 12 electrons we have the formula which, unfortunately, does not give the center carbon an octet of electrons. We must therefore move a pair of nonbonded electrons from the oxygen to form a new bond between carbon and oxygen. This move is shown below with an arrow

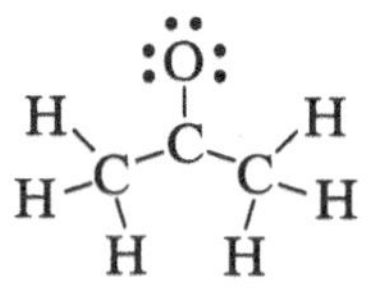

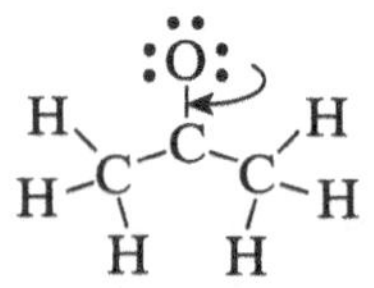

and produces a structure in which the octet rule is satisfied.

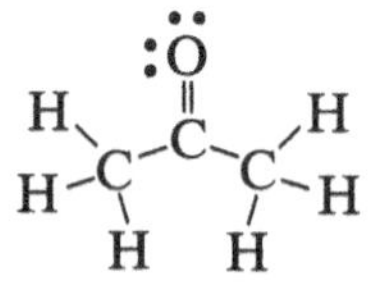

The bond between the carbon and the oxygen is a double bond. We have seen this very important group of atoms (C=O), namely, the carbonyl group, in ketones, aldehydes, acids, esters, acid halides, amides, and anhydrides in Chapter 2. ∎

Before we continue by discussing the formate ion, we should note that this is our first encounter with a method used to show how to rearrange electrons. The method, often called "*electron-pushing*," uses arrows to show the direction of electron movement. This is an artificial means of keeping track of electrons. The method assumes that the electrons are particles (as does the entire Lewis model), whereas we know that the more sophisticated wave model is required to explain many of the properties of electrons.

Statement. The head of the arrow always points directly to the new position of the electrons, and the tail of the arrow always begins where the electrons currently reside. An arrow can originate at a bond, at a lone pair of electrons, or, in some cases, at a single electron. The electrons can flow to an atom or to an area between two atoms.

The formate ion provides a good example of both the Lewis procedure and a structural complication. We begin by writing the atom connectivities for the ion:

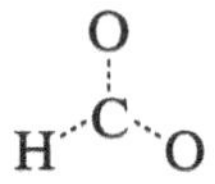

When we count the valence electrons, we must be sure to include the extra electron that makes this molecule an anion. The carbon atom has 4 valence electrons, each oxygen atom has 6, the hydrogen has 1, and the extra electron produces a total of 18 valence electrons. We distribute these 18 electrons about the atoms by first giving each set of attached atoms a pair of electrons, as represented by the lines in the following structure:

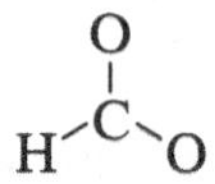

This leaves 12 more electrons, and we now give each of the oxygen atoms six electrons, arranged in pairs, so that each oxygen will now have a total of eight electrons (six electrons in the lone pairs plus the two electrons in the bond):

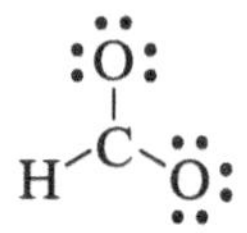

However, the carbon atom now has only six, not eight, electrons, and we must rearrange the electrons in order to obey the octet rule. In this type of situation it is usually possible to take a pair of electrons from one of the terminal atoms (an atom with one attached atom is a terminal atom) and convert it to a bonding pair. This rearrangement of electrons is shown with an arrow going from the nonbonding pair of electrons to the location of the new bond:

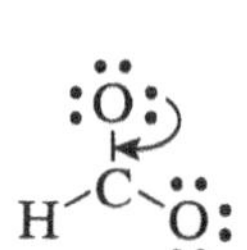

The new arrangement of electrons obeys the octet rule and can be considered a good Lewis structure:

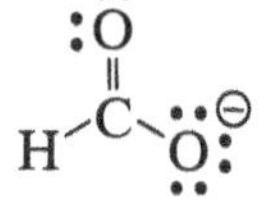

With this final structure the negative charge has been assigned to the oxygen with three lone (or nonbonding) pairs of electrons.

We now ask whether the Lewis structure above helps us predict any of the molecular parameters of the formate ion. One of the first things that we notice is that the two linkages between the carbon atom and the oxygen atoms are not the same. One of the attachments involves a double bond, whereas the other one is a single bond.

This circumstance can be interpreted as meaning that one of the bonds is stronger than the other because there are twice as many electrons between the nuclei and therefore a greater number of electron–nucleus attractions. Or, to use a simple analogy, if we have three pencils lying side by side with two of them attached by two rubber bands and the other two attached by only one rubber band, we can easily imagine that the pencils attached by two bands will be more strongly attached. In the case of atoms a double bond is stronger than a single bond and the atoms are therefore closer to one another—that is, the bond length is shorter. We can generalize this conclusion by the following statement.

Statement. The greater the number of bonds (this is also called the *bond order*), the stronger the bond and the shorter the bond.

Our Lewis formula for the formate ion therefore indicates two different CO bond lengths in the ion, one shorter than the other. However, the bond lengths in the formate ion have been measured by X-ray diffraction, and the lengths are identical. Clearly, the Lewis structure, even though it obeys the octet rule and has the correct number of valence electrons, does not provide a satisfactory description of the electronic structure—the bonding—in the formate ion.

Resonance

In order to remedy the deficiency in the description of the formate ion we resort to a ploy introduced by Linus Pauling (1901–1994), one of the great American chemists. This ploy, called *resonance hybridization*, recognizes that our decision to give the vertical CO linkage a double bond was arbitrary; that is, we cannot justify placement of the double bond to one oxygen rather than the other. So, as Pauling suggested, we can write another Lewis structure that contains the double bond to the other oxygen and then mix the two structures to form one hybrid structure. This hybridization process is frequently used in chemistry, and in reality is just the mathematical combination of structures to produce a new structure that better describes a molecule. The process is somewhat similar to mixing two cans of paint—one red and one white. When the paint is mixed, the resulting color is pink. The new color is a composite of the two original colors. Both the red and the white contribute equally and simultaneously to the new color.

Looking at the process a bit differently, the resonance hybrid is simply the composite of the two contributing Lewis structures. As an average, the number of bonds on both sides of the resonance hybrid is the same. The process of combining two individual Lewis structures to create a new hybrid structure is always depicted by using a double-headed arrow between the two contributing structures as shown below:

The Lewis structures that are hybridized are referred to as either *resonance forms*, *resonance structures*, or *resonance contributors*. The hybrid generally has a lower energy and different bond lengths compared to the individual resonance contributors. The hybrid should not be regarded as the "correct" electronic structure or a "correct" representation of the bonding; it is simply the "best" representation possible using Lewis structures.

Q Which of the following is not a resonance form for the formate ion?

A The third (rightmost) one is not a resonance structure of the formate ion because the hydrogen atom has been moved to bond with a different atom. Resonance forms always maintain the same atomic connectivity.

We have now produced a description of the bonding in the formate ion that gives equal bond lengths to the carbon–oxygen bonds. To determine whether these lengths agree with the experimentally determined CO lengths, we need to determine the bond order between the carbon and each of the oxygen atoms in the resonance hybrid. This bond order can be calculated as the average of the bonds between the atoms in each contributing structure. For each CO linkage there is one bond in one contributing structure and two bonds in the other structure, for an average of $\frac{3}{2} = 1.5$. Each linkage in the resonance hybrid therefore has the same bond order of 1.5.

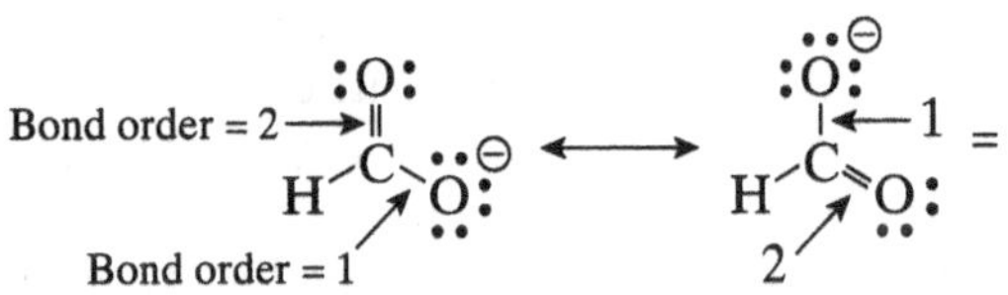

38 BONDING

Resonance forms

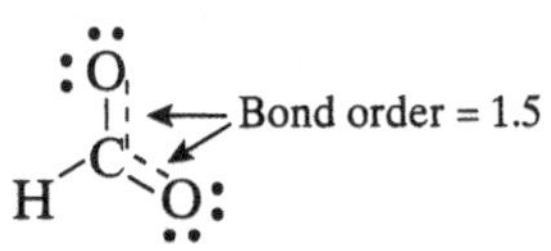

Depiction of resonance hybrid

The hybrid predicts equal bond lengths and, moreover, predicts a bond length between that of the CO single bond and that of the CO double bond.

Q The formate ion has two CO linkages, each of which has a length of 1.26 Å. The average CO single-bond length is 1.43 Å, and the average CO double-bond length is 1.23 Å. Do the resonance structures above rationalize the experimental bond length in the formate ion?

A Resonance hybridization gives a bond order of 1.5, so the length of the bond should be somewhere between that of a bond with a bond order of 1 (a single bond) and that of a bond with a bond order of 2 (a double bond). Thus, the bond length can be predicted to be about 1.3 Å (between the single- and double-bond lengths), which is reasonably close to the experimentally observed length of 1.26 Å. ∎

Resonance hybridization does not always give a good account of bond lengths. For example, in the nitrate ion the following three formulas can be hybridized to provide a composite structure, the resonance hybrid, which predicts equal bond lengths for each of the NO bonds.

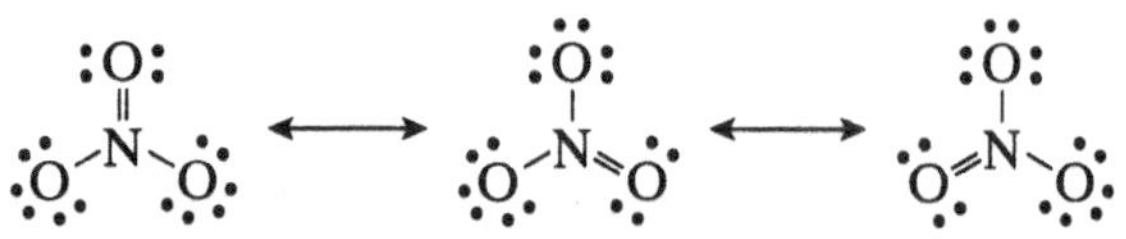

Q Give the bond order for each NO linkage and then predict the bond length in the nitrate ion.

A The bond order is $\frac{4}{3}$. If you have difficulty seeing this, take just the top linkage in each one of the resonance contributors and count up the number of bonds— 2 (the left form) + 1 (the middle form) + 1 (the right form) = 4. *Three* resonance forms contribute these *four* bonds, and therefore the average number of bonds in the composite is $\frac{4}{3}$. Since the bond order is $\frac{4}{3}$ we would expect the bond length to be just slightly shorter than that of a single bond, which we know from the data above is 1.43 Å. Therefore, we might guess that the resonance hybrid would predict a bond length of roughly 1.35 Å, but certainly not as short as 1.20 Å, the distance for an NO double bond.

Ah, but here is the rub: The experimental bond length of each NO bond in the nitrate ion is 1.22 Å, which is nearly identical to the NO double bond length. It appears that the resonance hybrid does not give a good description of the magnitude of the bond lengths in this ion. In the next section on formal charge we will find out how to make our estimate a bit better. ∎

Formal Charge

In cases such as the nitrate ion with its unexpectedly short bond lengths, it is helpful to assess the amount of electron density on each atom in the molecule with the objective of modifying slightly the prediction of bond length. The modification makes use of the notion that if two adjacent atoms have a considerable difference in electron density, one being deficient in electron density and the other having excess electron density, there will be an attraction between the two atoms that is very similar to the attraction between two ions. The atoms are, of course, not ions for they are held together by a covalent bond, but this bond can be said to have some ionic character. Unfortunately, it is difficult to estimate the amount of excess (or deficit of) electron density on each atom. These calculations can be made using molecular orbital theory or through the simple use of electronegativity differences, but we generally make use of a simple, albeit artificial, method called *formal charge*. The formal charge of an atom is the excess electron density on the atom when the electrons surrounding the atom are counted in a particular way. The method for calculating the electrons is as follows.

Statement. Count half of all the electrons shared with other atoms and all the electrons that belong only to the atom [nonbonding (NB) electrons]. Subtraction of this number from the number of valence electrons (VEs) for the neutral atom gives the formal charge on the atom:

$$\text{Formal charge} = \text{VE} - (\tfrac{1}{2}\,\text{shared} + \text{NB})$$

Consider the nitrogen atom in any one of the resonance forms of the nitrate ion.

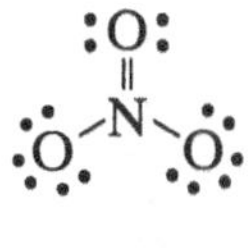

$$\text{:O:} \atop \overset{\|}{\underset{\text{:O:}}{\text{N}}}\!-\!\text{:O:}$$

There are four bonds to this atom, each representing two electrons, so we must count half of these eight shared electrons. There are no nonbonding electron pairs on the nitrogen, and thus the nitrogen, in the formalism, "sees" four valence electrons surrounding it. This number must be subtracted from the number of valence electrons required for nitrogen. The formal charge of the nitrogen is then $5 - 4 = 1$. The positive formal charge means that the nitrogen is *deficient* one electron. For each of the singly bonded oxygen atoms in the formula the formal charge is $6 - (\frac{2}{2} + 6) = -1$. The doubly bonded oxygen has a formal charge of $6 - (\frac{4}{2} + 4) = 0$. Formal charges are usually enclosed in circles to distinguish them from a real charge. Although we show this below, many organic texts do not follow this rule.

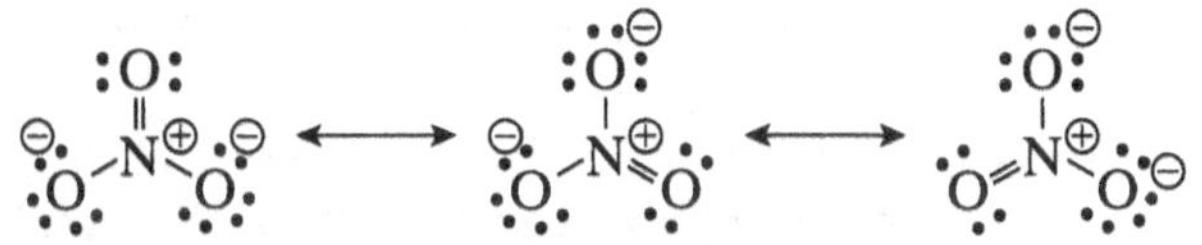

$$\overset{\ominus}{\text{:O:}}\!-\!\overset{\oplus}{\text{N}}\!=\!\overset{\ominus}{\text{:O:}} \longleftrightarrow \overset{\text{:O:}^{\ominus}}{\underset{\text{:O:}}{\overset{\|}{\text{N}^{\oplus}}}}\!-\!\text{:O:} \longleftrightarrow \overset{\ominus}{\text{:O:}}\!-\!\overset{\oplus}{\text{N}}\!=\!\text{:O:}^{\ominus}$$

In the resonance hybrid each oxygen atom has a formal charge of $-\frac{2}{3}$. In other words, two electrons are spread out over three oxygen atoms. The importance of this analysis is that the electron-deficient nitrogen atom is surrounded by electron-rich oxygen atoms and the NO bond consequently has some ionic character. Because of these adjacent opposite "charges," the nitrogen and attached oxygens are attracted slightly to one another, and this makes the bond shorter than expected from consideration of the bond order alone.

Q Determine the average formal charge on each oxygen atom of the formate ion.

A

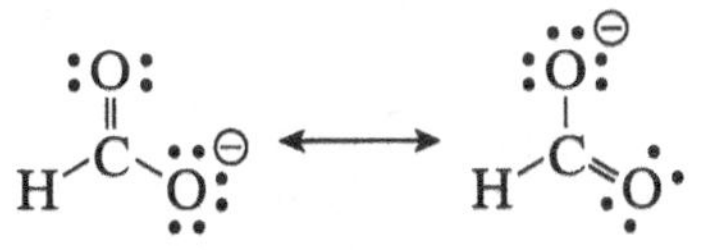

$$\text{H}\!-\!\overset{\overset{\text{:O:}}{\|}}{\text{C}}\!-\!\text{:O:}^{\ominus} \longleftrightarrow \text{H}\!-\!\overset{\overset{\text{:O:}^{\ominus}}{|}}{\text{C}}\!=\!\text{:O:}$$

In these resonance structures we have enclosed the formal negative charge for the singly bonded oxygen in a circle. The formal charges on all other atoms are zero. It is customary not to show zero formal charges. Note that the sum of all of the formal charges must add up to the charge on the species. The average formal charge on each oxygen is $-\frac{1}{2}$. ∎

We can begin to recognize the formal charges on atoms in certain bonding situations. For example, an oxygen with two pairs of nonbonded electrons will always have a zero formal charge while an oxygen with one nonbonded pair of electrons will always have a +1 formal charge. Nitrogen with one lone pair will have a zero formal charge, whereas a nitrogen with no lone pair will have a +1 formal charge. Carbon with no lone pairs will always have a zero formal charge, while a carbon with one lone pair will have a −1 formal charge. These situations are illustrated below.

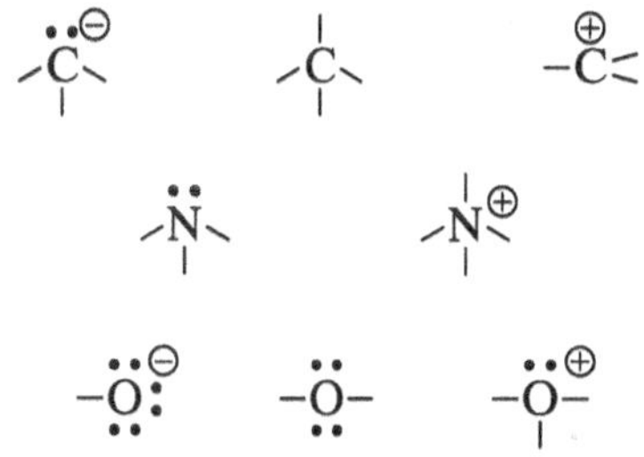

Q Write a Lewis structure for benzene, which, as you know, has the molecular formula C_6H_6 and has a structural formula that indicates that each hydrogen atom is attached to one carbon atom. The carbon atoms are attached to one another in a ring.

A

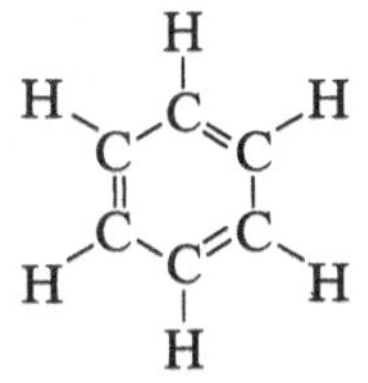

$$\begin{array}{c}\text{H}\\|\\\text{C}\end{array}$$

(benzene Kekulé structure with alternating double bonds and H on each carbon)

You should recognize that this structure predicts that benzene has three carbon–carbon bonds that are shorter than the other three; in other words, the three double bonds should be shorter than the three single bonds. ∎

Q The experimental bond lengths of benzene, however, are all 1.40 Å. The average or typical carbon–carbon single-bond length is 1.54 Å and the double-bond length is 1.33 Å, which implies that the carbon–carbon bonds in benzene have a bond order between those of a single bond and a double bond. Write resonance forms that rationalize this observation.

40 BONDING

A By simply changing the positions of the alternate double bonds, we can write two resonance forms.

A hybrid of these two forms has a bond order of 1.5, which agrees with the experimental data. This hybrid cannot be illustrated by one Lewis formula alone, and most organic chemists use the structure

with the understanding that all carbon–carbon bonds lengths are the same as a result of resonance. The resonance hybrid description of benzene also explains the observations that the carbon–carbon bonds in benzene have characteristics of neither a single bond nor a double bond, and that, although not obvious at the moment, the electrons are delocalized or spread over the six carbon–carbon bonds. ∎

Generating Resonance Structures by Using Electron Flow ("Pushing Electrons")

This last characteristic can perhaps be better appreciated by looking at a way to generate the two hybrid structures. Let's assume that we have just written the Lewis structure for benzene and want to generate another resonance structure.

Q Move two of the four electrons in a double bond to one of the carbons of the double bond. Use an arrow to show how the electrons flow.

A Each line between each set of adjacent carbons represents two electrons. An arrow is used to show the movement of the two electrons in the double bond between carbons 1 and 2 to carbon 2.

∎

Q Determine the formal charge on carbons 1 and 2 after the electrons have been moved.

A This leaves carbon 1 with three bonds and a formal charge of $+1 (4 - \frac{6}{2} = 1)$.

The electron pair that previously rendered carbon 1 neutral and was shared by carbons 1 and 2 is now located on carbon 2. Thus, around carbon 2 there are three bonds and two nonbonding electrons for a formal charge of $4 - (\frac{6}{2} + 2) = -1$. ∎

Q Next, move the two nonbonding electrons on carbon 2 into the region between carbons 2 and 3, thereby creating a double bond between carbons 2 and 3.

A In order to form the double bond between carbons 2 and 3, the double bond already existing between carbons 3 and 4 must simultaneously be broken. If it were not broken, carbon 3 would have five bonds, a clear violation of the octet rule. As the bond between carbons 3 and 4 is broken, the pair of electrons forming the bond moves to carbon 4, giving it two nonbonding electrons and a formal charge of −1.

This process is repeated until finally a double bond is formed between carbons 6 and 1, thereby giving carbon 1 a zero formal charge. The overall process is represented as simultaneous movement of three pairs of electrons in double bonds with arrows showing the direction in which the electrons move (see below). It is very important to position these arrows correctly. ∎

Statement. When the arrow extends from the center of the bond to the adjacent atom, this indicates that the two electrons represented by the bond (the line) wind up on the atom to which the arrow points. When the arrow extends from the center of a double bond to the center of another bond, a pair of electrons creates a new double bond.

In any event the result of the various electronic movements produces the other resonance structure.

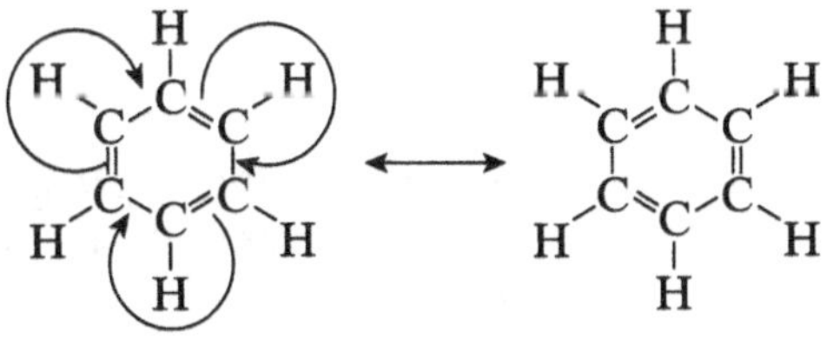

[Resonance structures of benzene with electron-pushing arrows]

This electron-pushing procedure can be used to quickly create resonance forms for a great variety of molecules.

Q Write a Lewis structure for formamide, $HCONH_2$.

A

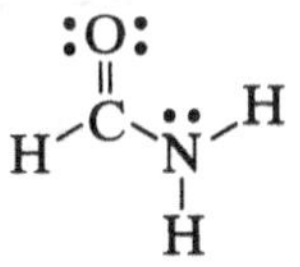

[Lewis structure of formamide: H-C(=O)-NH_2]

Q Use electron pushing to move the nonbonded pair on the nitrogen into the region between the carbon and the nitrogen. Then follow up by moving two electrons in the double bond between the carbon and the oxygen onto the oxygen.

A The scheme below shows the two arrows that are necessary to show the overall movement of electrons from the nitrogen to the oxygen. The resonance form that results has a formal positive charge on nitrogen and a negative formal charge on the oxygen.

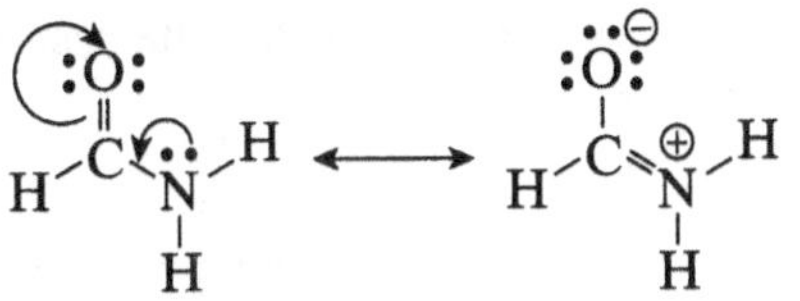

[Resonance structures of formamide showing electron pushing]

After writing two or more resonance forms we must always inquire about the importance of each form. In the case of amides, like formamide, there is good evidence, based on bond lengths and on bond energies, that the two resonance forms contribute almost equally to the overall electronic structure of the molecule.

Q The two resonance forms of the formate ion *must* contribute equally to the resonance hybrid. Why is this true?

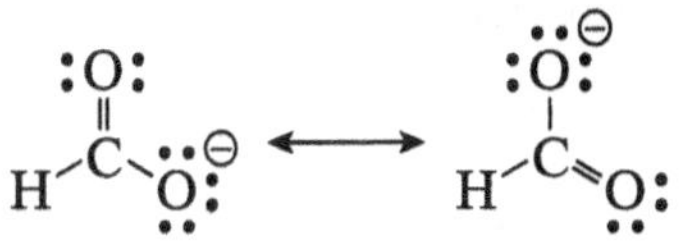

[Two resonance structures of formate ion]

A It must be true because the placement of the double bond (to either the "top" or "side" oxygen) is purely arbitrary. The ion is symmetric and, as we have seen, the CO bonds have exactly the same length. ■

Q Use electron pushing to write three resonance forms for methyl vinyl ketone.

A

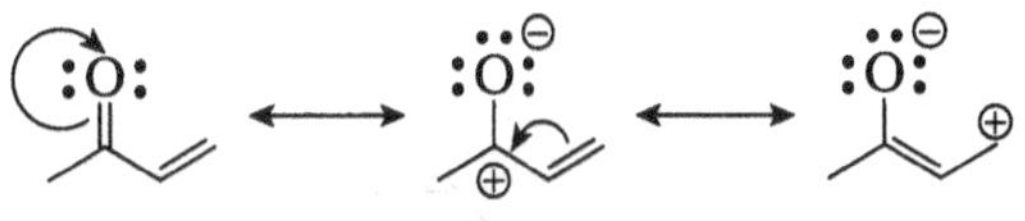

[Three resonance structures (a), (b), (c) of methyl vinyl ketone]

(a)   (b)   (c)

Note that resonance form **a** was used to generate form **b**, which was then used to produce form **c**. ■

Q In the case of the resonance forms above, there is no reason based on symmetry considerations to believe that the two resonance forms must contribute equally to the resonance hybrid. If the carbon–carbon bond in **a** above (the vinyl group) has a length just a bit longer than that of a typical C=C bond, which structure **b** or **c** would contribute more?

A Structure **b** above, which has the C=C bond. ■

42 BONDING

Next we write resonance structures for the anion formed by acetaldehyde (ethanal). The Lewis structure of ethanal is

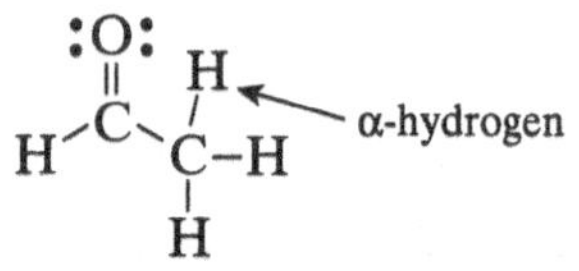

$$\underset{H}{\overset{:\ddot{O}:}{\underset{}{\overset{\parallel}{C}}}}\!\!-\!\!\underset{\underset{H}{|}}{\overset{H}{\underset{|}{C}}}\!\!-\!\!H \quad \leftarrow \alpha\text{-hydrogen}$$

When ethanal reacts with hydroxide ion, a hydrogen ion is removed by the hydroxide ion from the carbon adjacent (alpha) to the carbonyl group in an acid–base reaction. The electron withdrawing carbonyl group makes this hydrogen slightly acidic.

Q Write a Lewis structure for the anion formed, and then use electron pushing to write another resonance structure of the anion.

A

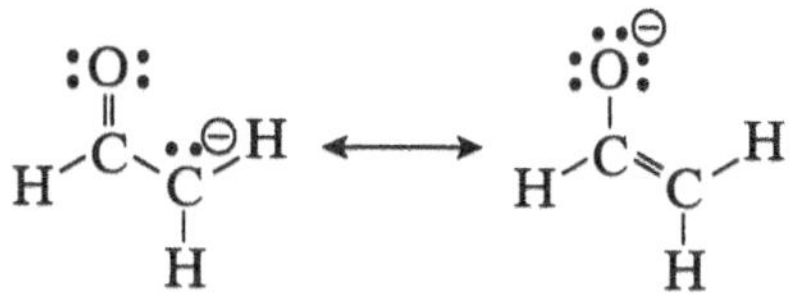

[resonance structures of the ethanal anion]

Note that the second form was created using the electron-pushing scheme:

[electron-pushing scheme diagram]

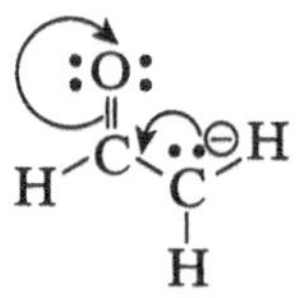

The delocalization of electron density as shown by the resonance structures lowers the energy of the anion. This lower-than-expected energy is part of the reason why ethanal is slightly acidic. ■

Statement. Generally, delocalization of electrons lowers the potential energy of a molecule or ion and therefore stabilizes it.

The amount by which the energy of a substance is lowered by virtue of electron delocalization is called its *resonance energy*. This energy is relative to the energy of one of the resonance forms. For benzene the resonance energy is believed to be approximately 150 kJ/mol.

Q Which of the two resonance structures below is more important for acetone?

[two resonance structures of acetone labeled (a) and (b)]

A Structure **a**, showing the carbonyl group, is more important because it obeys the octet rule and does not have charge separation associated with it. ■

If you have been watching carefully, you may have noticed that the resonance structures for the anion of ethanal and those for formamide have the same forms. It is also true that ethanal and formamide have the same number of valence electrons; that is, they are *isoelectronic*.

Statement. It is generally true that isoelectronic species have similar Lewis structures.

Q Try your hand at recognizing isoelectronic groups by choosing the isoelectronic pairs in the following set of compounds:

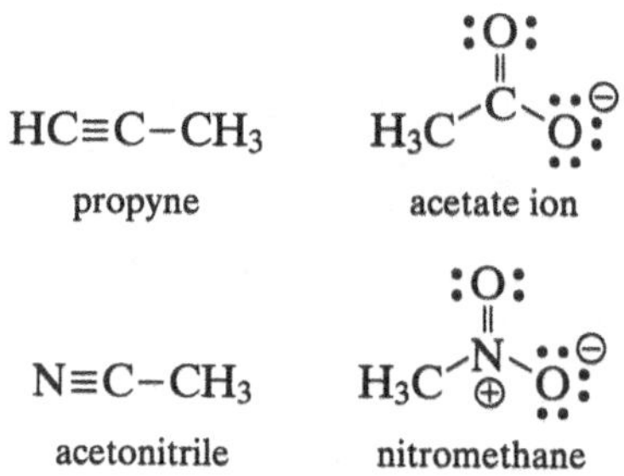

$HC\equiv C-CH_3$ [structure of acetate ion]
propyne   acetate ion

$N\equiv C-CH_3$ [structure of nitromethane]
acetonitrile   nitromethane

A The acetate ion and nitromethane are isoelectronic. Propyne and acetonitrile are isoelectronic. ■

For our next example, we will write resonance structures for methoxybenzene.

Q Write a Lewis structure for methoxybenzene and provide an alternative name.

A

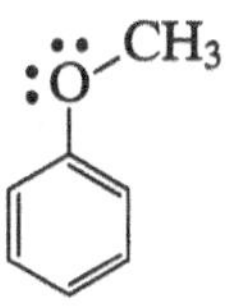

The compound can also be named methyl phenyl ether and has the trivial (but frequently used) name of anisole. ■

In the conventional Lewis structure for this molecule there are two lone pairs of electrons on the oxygen of the methoxy group. In order to generate other structures we start with a lone pair on the oxygen atom and move electron pairs as indicated by the arrows below.

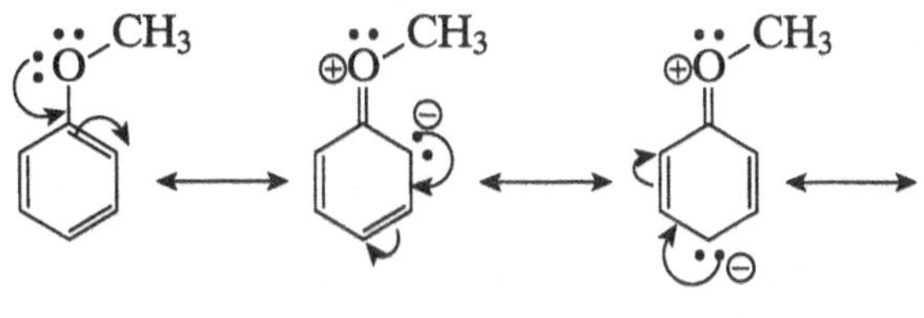

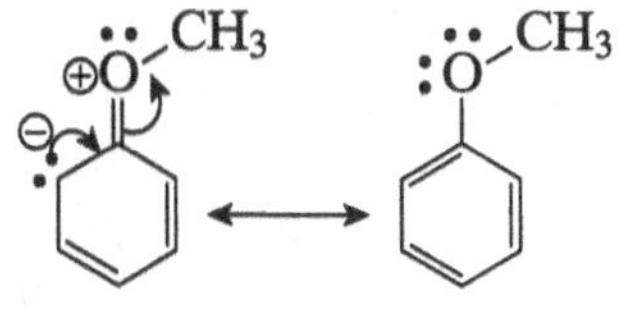

The movement shown by arrows on structure **a** puts a lone pair on one of the carbons in the benzene ring. Continued movement as shown on structure **b** places a pair of electrons (and the negative charge) on the position opposite the oxygen atom. Similar movements are shown in structures **c** and **d**. The final movement generates a structure of methoxybenzene that represents resonance in the benzene ring. (Compare the positions of the double bonds in structures **a** and **e**.) It is also noteworthy that the methoxy group places additional electron density on the two *ortho* positions and the *para* position of the benzene ring, at least according to our resonance forms. This redistribution of electron density caused by substituents will be important when you study the way in which reagents attach to the benzene ring.

Exceptions to the Octet Rule

Thus far, we have created primarily resonance structures that obey the octet rule. There are exceptions to the octet rule that occur rather frequently in compounds that contain elements below the second period. The bonding in phosphorus pentafluoride is normally described with the structure shown below, with phosphorus having five bonds and therefore a total of 10 electrons in its valence shell.

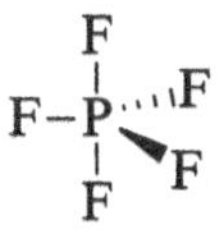

This violation of the rule of eight is possible and permissible for elements that have more than just the *s* and *p* orbitals in their valence shells. In the valence shell of phosphorus there are *s*, *p*, and *d* orbitals, and these orbitals can hold more than eight electrons.

Because most organic compounds contain only first- and second-period elements such as H, C, O, and N, expansion of the octet is not possible for these compounds. *Organic (and other) compounds can, however, have resonance structures that have atoms with fewer than eight electrons.* The conventional Lewis structure for methyl vinyl ketone is shown again below along with two possible resonance forms that violate the octet rule. Both **b** and **c** contain a carbon with a formal positive charge; it has only three bonds and therefore only six electrons.

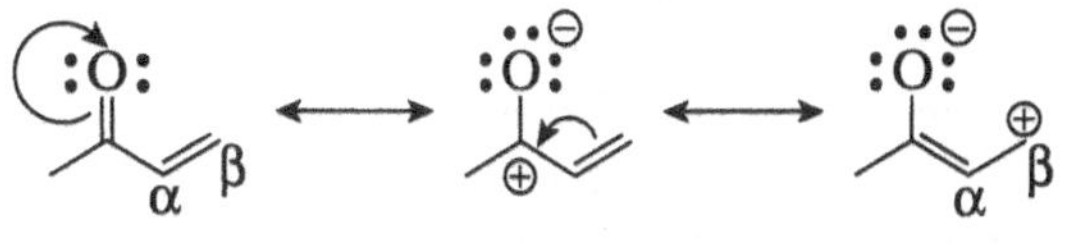

Q   Draw a resonance structure for each of the following: nitromethane, cyanobenzene (benzonitrile), and the anion of acetic acid (acetate ion).

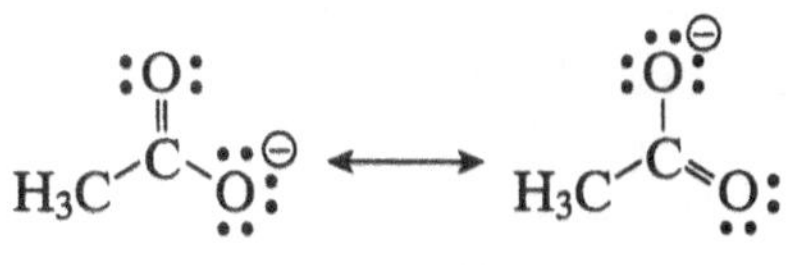

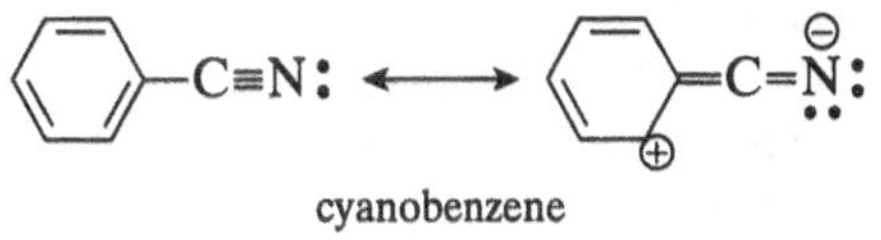

A

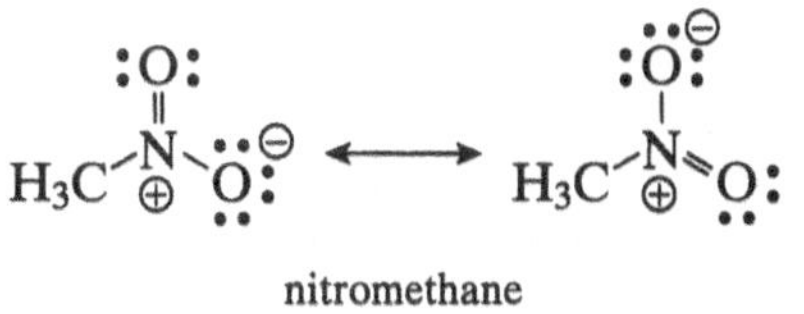

■

Q (1) Which of the compounds shown above must have resonance structures that contribute equally to the resonance hybrid? (2) For which compound is there a resonance structure that does not obey the octet rule? (3) Which species are isoelectronic? (4) Which has the shortest bond between the central atom and oxygen—nitromethane or the acetate ion? (5) Give an alternative name for cyanobenzene and the acetate ion.

A (1) Nitromethane and the acetate ion due to symmetry. (2) Cyanobenzene. (3) Nitromethane and the acetate ion. (4) Nitromethane, due to the attraction of the opposite formal charges. (5) Benzonitrile and the ethanoate ion. ■

THE VALENCE BOND MODEL

During the development of wave mechanics and the recognition that electrons can be described by mathematical constructs known as *orbitals*, it was important to produce an orbital equivalent of the Lewis model. In the valence bond model it is assumed that bonds are formed by the overlap (mathematical addition) of orbitals on adjacent atoms. For example, the bond between the two fluorine atoms in diatomic fluorine (F_2) is assumed to be formed by the overlap of two p orbitals (see below).

By convention the F–F bond is assumed to lie on the x axis, and therefore, only the p_x orbitals can overlap one another.

Q You are probably wondering whether two p_y (or two p_z) orbitals could overlap to form the F–F bond.

A Two p_y orbitals have the correct orientation to overlap, as do two p_z orbitals. Overlap of these orbitals forms a different kind of bond. As shown below, this overlap produces electron density above and below the line connecting the two fluorine atoms.

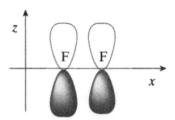

Overlap must occur between two parallel p orbitals. Note that the symbol for fluorine designates the position of the nucleus and is shown above the x axis only to make it visible. ■

Overlap of orbitals, similar to the constructive interference of two waves, leads to the buildup of electron density between the nuclei. Bonds that have electron density on the internuclear (x-axis) line are known as *sigma bonds*. In diatomic oxygen a sigma bond is formed by overlap of p_x orbitals, and the second bond between the oxygen atoms, as shown in the Lewis structure

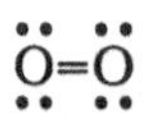

$$\ddot{\underset{..}{O}}=\ddot{\underset{..}{O}}$$

is produced by the overlap of p orbitals that are perpendicular to the x axis. These orbitals could be either the p_z or the p_y (Generally the z axis is assumed to be vertical; i.e., in the plane of the paper, while the y axis protrudes from the plane of the paper; that is, toward the reader.) If the p_z is used by one oxygen atom, then the other oxygen must also use a p_z orbital to form this bond because a p_z orbital and a p_y orbital will not overlap.

Although there is no electron density on the internuclear axis for a bond created by overlap of two p_z (p_y) orbitals, there is sufficient density between the nuclei above and below the x axis to result in considerable stabilization of the molecule. This orbital (and its p_y counterpart) is called a *pi bond*.

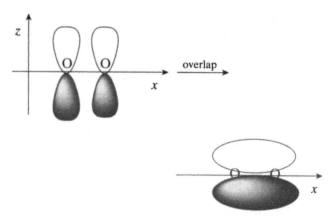

The elliptical regions on the right indicate areas of relatively high electron density; only two electrons, with their spins opposed, occupy this region (orbital).

A pi (π) bond is always somewhat less stabilizing than a sigma (σ) bond, which means that π bonds are rarely found without a σ bond. The valence bond model therefore portrays a double bond as a combination of a σ bond and a π bond.

Triatomic Molecules

At first glance, the application of the valence bond model to a triatomic molecule such as water should be simple. The oxygen atom in water has one s orbital and three p orbitals, and therefore the following scenarios could be imagined: (1) an s orbital and a p orbital on oxygen overlap with the s orbitals of two hydrogens, and (2) two p orbitals on oxygen overlap with the two hydrogen s orbitals.

Q If one hydrogen atom's s orbital overlaps with an s orbital on oxygen, whereas the other hydrogen atom's s orbital overlaps with a p orbital on the oxygen (see diagram below), would the two O–H bonds have the same bond energies?

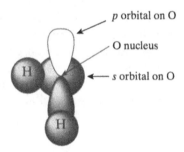

Overlap of s and p orbitals on oxygen with the s orbital on each hydrogen.

A The s and p orbitals have different energies as shown in the diagram below. One consequence of this circumstance is that the stability of a bond formed by overlap of an s orbital on one atom with an s orbital on another atom must be different from the stability of a bond formed by overlap of a p orbital with an s orbital.

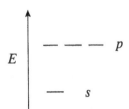

Relative energies of s orbitals and p orbitals.

In scenario 1, overlap of an s orbital on oxygen with an s orbital on hydrogen should produce a bond with a lower energy (because of the lower energy of a 2s orbital compared to a 2p orbital) than the bond produced by overlap of a p orbital on oxygen with an s orbital on hydrogen. ∎

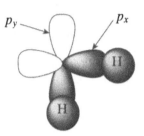

Overlap of two oxygen p orbitals with the s orbital of each hydrogen.

Scenario 2, overlap of two p orbitals on oxygen, as shown above should produce bonds that have the same energy and length because the two p orbitals on oxygen have identical energies and shapes (but not identical orientations). Because the two p orbitals used by the oxygen must be directed at 90° to one another, we can predict that the bond angle adopted by a molecule that would use this bonding would be close to 90° (see above).

Q Can you explain why these bonds must be oriented at 90° to one another?

A The bonds must be oriented at 90° to one another because they are formed by p orbitals (the p_y and p_z) that are directed along the y and z axes of the Cartesian coordinate system. The p_y and p_z (and p_x) orbitals are oriented at 90° to one another. ∎

As always, the consequences of each model must be compared with the experimental facts about the molecule. Water has two OH bonds that have the same length and strength and a bond angle of 105°. Scenario 1 therefore does not agree with the experimental data because the bond formed by overlap of an s with an s orbital does not produce a bond of the same strength and length as the bond formed by overlap of an s orbital with a p orbital.

Scenario 2 also does not agree well with the experimental parameters for water—the actual bond angle is 105°, not 90°. Although other factors affect the bond angle, such as repulsion of the electron density

46 BONDING

surrounding the hydrogen atoms, theoretical chemists have established that these other factors probably cannot account for the difference in bond angle.

Orbital Hybridization

Linus Pauling came to the rescue in this dilemma by explaining that orbitals are merely mathematical constructs and can be manipulated to produce other entirely acceptable orbitals. These newly created orbitals are known as *hybrid orbitals* because they are constructed by mixing the *s* and *p* and, for some elements, *d* orbitals on an atom. Although there are many possible combinations of *s*, *p*, and *d* orbitals, the combinations that are of importance in organic chemistry are as shown below:

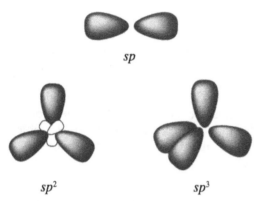

Hybrid orbitals for organic molecules. (*Note*: The smaller lobes of the *sp* and sp^3 orbitals are obscured by the larger lobes.) Each orbital consists of a large lobe (shaded dark) and a small backend lobe (not shaded).

Most organic molecules contain atoms such as C, N, and O that do not have *d* orbitals in their valence shells. Consequently, we rarely need to use hybrids that involve *d* orbitals.

The first set of hybrid orbitals, the *sp* hybrids, are directed toward the opposite ends of a straight line. The angle between these orbitals is 180°. In every other respect, such as their energy and their extension in space (but not their orientation), these orbitals are identical. The second set of orbitals, the sp^2 set, is formed by mixing one *s* orbital and two *p* orbitals. These orbitals are located in a single plane and are directed toward the corners of an equilateral triangle. The angle between adjacent orbitals is 120°. The third set is a mixture of one *s* orbital and three *p* orbitals. These orbitals are directed toward the corners of a tetrahedron with angles between them of 109°. The relationship between the type of hybridization and bond angles is summarized in Table 3.1.

TABLE 3.1. Hybridization–Bond Angle Relationship

Hybridization	Bond Angle
sp	180°
sp^2	120°
sp^3	109.5°

 Here again is that fantastic geometric object, the tetrahedron. Draw it and place a carbon in the center that has sp^3 hybridization.

A

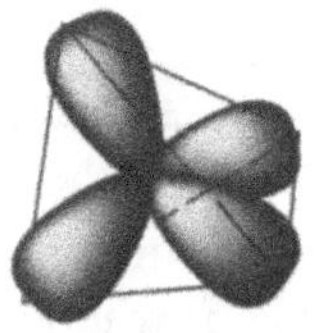

In order to refine our valence bond description of the bonding in water, we now choose the set of hybrid orbitals that best matches the actual bond angle in water (105°). The set of sp^3 hybrid orbitals has an angle of 109°, which is closer than the angles of the *sp* and sp^2 hybrid orbitals to the experimental angle in water. Consequently, we choose sp^3 hybrid orbitals to overlap the *s* orbitals on the hydrogen atoms. We can picture the bonding as shown below. Note that the lone pair electrons are also in sp^3 hybrid orbitals.

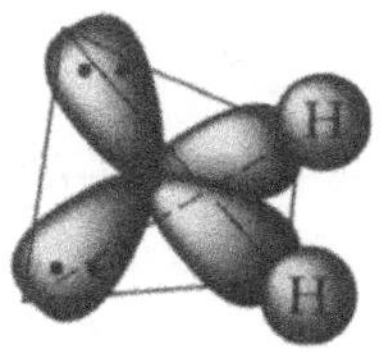

Use of sp^3 hybrid orbitals for bonding in water.

Our next example of the application of the valence bond model is ethene, for which the Lewis formula is

A

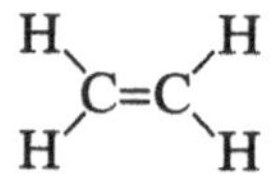

$$\begin{array}{c} H \\ \diagdown \\ C=C \\ \diagup \\ H \end{array} \begin{array}{c} H \\ \diagup \\ \\ \diagdown \\ H \end{array}$$

 What do we need to know about this molecule before we can describe it with the valence bond model?

A In order to choose the best set of orbitals, we need to know that the H–C–C bond angles (and H–C–H angles) are about 120°. This angle calls for the use of sp^2 hybrid orbitals (see Table 3.1). These orbitals must be used for the formation of σ bonds. (σ bonds and lone pair electrons determine the geometry of a molecule, as we will see later). ∎

The σ-bond framework is shown below, where the C–C σ bond is formed by overlap of two sp^2 hybrid orbitals, one on each of the adjacent carbons, and the C–H bonds are formed by overlap of sp^2 hybrids with the hydrogen s orbitals.

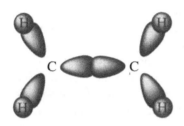

Use of sp^3 hybrids in formation of σ-bond framework in ethene.

The Lewis structure shows that ethene also contains another bond between the two carbons. This bond must be formed by overlap of the p orbital that remains on each carbon. (Remember that sp^2 hybrids involve mixing one s orbital and two p orbitals and that therefore one p orbital on the carbon remains available.) The sp^2 hybrid orbitals lie in a plane that we will designate the xy plane. The hybrids must be formed by mixing an s orbital with a p_x orbital and a p_y orbital (use of a p_z would not produce orbitals in the xy plane). The pure p orbital on each carbon is therefore a p_z orbital. The two p_z orbitals overlap in π fashion to establish the double bond. The whole bonding scheme is shown below and predicts that because both carbons have sp^2 hybrids in the xy plane, all of the atoms must lie in the xy plane. Ethene is consequently predicted to be a planar molecule, and this prediction is borne out experimentally.

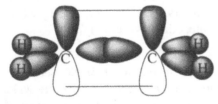

Both σ and π bonds in ethene. The π bond, shown by the lines connecting the two p orbitals, is perpendicular to the σ-bond framework. (*Note*: The smaller lobes of the sp^2 orbitals are obscured by the larger ones.)

Q This description of the bonding in ethene nicely accounts for the lack of rotation of the groups at a π bond. Can you explain this?

A The bonding in a *cis*-alkene is shown below, along with the consequences of the rotation of the right-hand group around the double bond. This rotation can only occur if the π bond is disrupted by eliminating the overlap between the two p_z orbitals. Considerable energy is required to break this bond, and therefore the groups attached to each end of the C=C bond are rigid at room temperature. We usually express this by saying that there is no free rotation around the double bond.

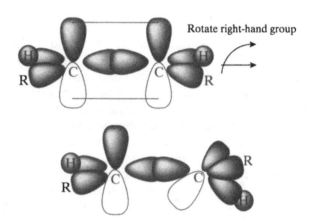

Rotation of the CHR group on the right around the C–C σ bond disrupts the π overlap. ∎

Q Methanal (formaldehyde) also utilizes an sp^2 hybrid on the carbon for the σ bonds. The π bond shown in the Lewis structure (note that a Lewis structure always accompanies a valence bond description) is formed by overlap of a p_z orbital on carbon with a p_z orbital on oxygen. Some instructors use hybridization on the oxygen as well, but there is no need to do so. There is no bond angle at oxygen because there is only one atom attached to the oxygen and therefore there is no experimental angle to describe using a hybrid. However, some instructors believe that the use of an sp^2 hybrid at oxygen for the formation of the σ bond also provides for the symmetric distribution of lone pairs at oxygen. In other words, the sp^2 hybrid orbitals on oxygen are used to house the σ bond to carbon and the two sets of lone pairs on the oxygen. In the Lewis structure this is sometimes stressed by placing the dots as shown below:

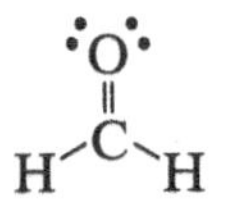

In this text we will assume that terminal atoms (like the oxygen in methanal) should not be hybridized. We will also generally distribute the lone pairs of electrons at oxygen as shown below:

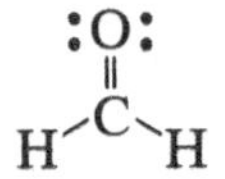

Does this description reveal anything about the geometry of the molecule?

A The geometry of any molecule is determined by its σ bonds. For methanal the sp^2 hybrid orbitals are chosen in order to produce a bond angle around the carbon of about 120°. Because our convention is to use the s, p_x, and p_y orbitals to generate the sp^2 hybrids, the carbon and both hydrogen atoms lie in the xy plane. The oxygen must also use an orbital in the xy plane (like the p_x) to most effectively overlap with the carbon's sp^2 hybrid orbital. Consequently, this model predicts that all of the atoms lie in the same plane, in other words, that the molecule is completely planar. ■

Our final example is the bonding in ethyne, HCCH.

Q If you don't remember that this molecule contains a carbon–carbon triple bond according to the Lewis model, the name provides that information. Alkynes have bond angles of 180° at both carbons of the triple bond, and this tells us that we must use sp hybrids in order to obtain those angles. What orbitals are used to form the bonds between the carbons and the hydrogens? What orbitals are used to form the triple bond?

A The σ-bond framework is shown below using sp hybrid orbitals constructed from an s orbital and a p_x orbital on each carbon. The p_y and p_z orbitals remaining on each carbon overlap to form two π bonds as shown below. The bond between a carbon and the attached hydrogen is formed by overlap of an sp hybrid on carbon with an s orbital on hydrogen.

Bonding in ethyne as depicted by the valence bond model. ■

Q Some organic reactions depend on the relative acidities of hydrogen atoms bonded to carbon. Generally the ease with which a hydrogen ion can be removed from another atom depends on the effective electronegativity of that atom—the greater the effective electronegativity, the easier it is to remove the hydrogen ion. Hydrogen is attached to the same atom (carbon) in ethane, ethene, and ethyne, but the hybrid orbitals used for the bonding differ in their relative energies. The diagram below shows the relative energies of the atomic orbitals and the hybrid orbitals.

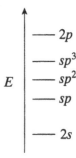

Relative energies of hybrid and atomic orbitals.

Because sp hybrid orbitals are composed of one s orbital and one p orbital, their energies are midway between those of s and p orbitals. Because the sp hybrids have the greatest percent s character, they have the lowest energy of any of the different types of hybrids. The lower energy of the electrons in s orbitals is a result of the fact that these electrons penetrate more closely to the nucleus. Electronegativity is a measure of the attraction of a molecule or atom for an electron, and therefore the sp hybrids have the highest effective electronegativity. Which is the most acidic hydrocarbon—ethane, ethene, or ethyne?

A Ethyne is the most acidic because of the sp hybridization used (according to the valence bond model) by the carbons. Ethyne, however, is seldom considered even a weak acid in water. It is only when ethyne is able to react with a strong base that it will lose a hydrogen ion. ■

THE VALENCE SHELL ELECTRON PAIR REPULSION MODEL

You may have noticed that carbons with four attached atoms, such as in ethane, generally have their bonds oriented toward the corners of a tetrahedron (approximately 109° bond angles) and are therefore said to use sp^3 hybridization. Carbons with three attached atoms, such as methanal, have σ bonds directed to the corners of a triangle with bond angles of ~120°. These carbons use sp^2 hybridization. Carbons with two attached atoms, such as in ethyne, have a linear geometry (180° bond angles) and use sp hybridization. This observation can be stated in the form of the *valence shell electron pair repulsion* (VSEPR) model.

Statement. The geometry of the linkages at an atom is determined by the repulsion of sigma-bonded and non-bonded electron pairs in the valence shell of that atom.

An atom always strives to minimize the repulsions between these pairs of electrons by having them oriented as far apart as possible. For methane with four bonded pairs of electrons, the repulsion between these pairs of σ-bonded electrons is minimized when the pairs are positioned at the corners of a tetrahedron. For methanal, the repulsion between three bonded pairs is minimized by placement at the corners of an equilateral triangle.

This model allows us to quickly discern the hybridization in almost any molecule. For example, in propyne

$$H-\overset{\overset{H}{|}}{\underset{\underset{H}{|}}{C}}-C\equiv C-H$$

the H–C–H angles of the CH₃ group are about 109° because there are four σ bonds at this carbon. At the other two carbons there are only two σ bonds and no lone pairs of electrons, and consequently these atoms should have the σ bonds directed along a straight line. More specifically, the C–C–C angle and the C–C–H angle should be approximately 180°, and these atoms therefore are said to use sp hybridization.

Q According to the VSEPR model, what orbitals are used to house the lone pairs on the oxygen of methanol?

A The VSEPR model, itself does not predict *hybridization*. It is used to predict the *geometry* around an atom. After that geometry is known, the valence bond model is used to specify the hybridization. Thus, at the oxygen in methanol there are two bonded pairs of electrons and two lone pairs of electrons, and the VSEPR model suggests that the repulsions between these four pairs of electrons are minimized by locating them at the corners of a tetrahedron. The C–O–H bond angle is therefore close to 109°. The valence bond model uses sp^3 hybridization to account for that bond angle. Therefore, the lone pairs of electrons, according to the *valence bond model*, are located in sp^3 hybrid orbitals. ■

Q Make a sketch to show the orbital overlap that creates only the σ bonds in ethyne.

A

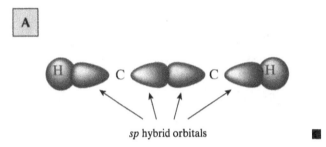

sp hybrid orbitals ■

Q Use the VSEPR model to specify the bond angles at each of the nonterminal atoms in acetamide.

A In acetamide

there are three nonterminal atoms—the carbon of the methyl group, the carbon of the carbonyl group, and the nitrogen of the NH₂ group. The methyl carbon has four σ bonds surrounding it and therefore should have bond angles of approximately 109°. The carbonyl carbon has three σ bonds and no lone pairs and therefore has bond angles of ~120°. The NH₂ nitrogen has three σ bonds and one lone pair (note that you must draw the electron dot formula to know that this nitrogen has a lone pair) and therefore has bond angles of about 109°. The valence bond model predicts that the hybridizations are sp^3 for the methyl carbon and the amine nitrogen and sp^2 for the carbonyl carbon. ■

Q Predict the hybridization at the non-terminal atoms in the following molecule:

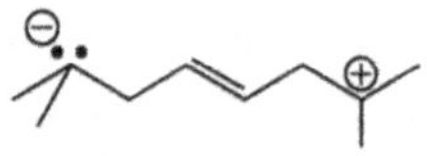

A This species contains a carbon with three σ bonds and a lone pair of electrons. This carbon has a formal negative charge and is referred to as a *carbanion*. At the other end of the molecule there is a carbon with three σ bonds and no lone pair of electrons. This carbon has a positive formal charge and is referred to as a *carbocation*.

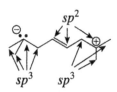

Q Predict the geometry of the carbanion and carbocation pictured below:

A The carbanion has carbons at three corners of a tetrahedron and a lone pair of electrons at the other corner. Therefore, the carbanion is pyramidal with bond angles of ~109°. The carbocation is trigonal planar with bond angles of ~120°.

4

STRUCTURE, ISOMERISM, AND STEREOCHEMISTRY

One important aspect of structure is *isomerism*, a word derived from the Greek terms *iso*, meaning the same or equal, and *mer*, meaning form, type, or part. Isomers have the same form or type of some aspect of their formulas. The two main types of isomers are *structural isomers* and *stereoisomers*. Structural isomers or constitutional isomers have the same molecular formula but different structural formulas, whereas stereoisomers have the same structural formulas, but different three-dimensional structures.

STRUCTURAL ISOMERS

Let's consider first a simple alkane, such as pentane, which has a molecular formula of C_5H_{12}.

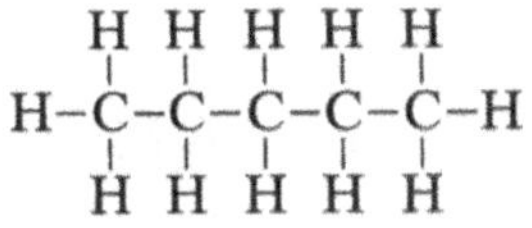

```
    H H H H H
    | | | | |
H - C-C-C-C-C - H
    | | | | |
    H H H H H
```

Q Provide two other ways to arrange five carbons and maintain the same molecular formula.

The Bridge to Organic Chemistry: Concepts and Nomenclature
By Claude H. Yoder, Phyllis A. Leber, and Marcus W. Thomsen
Copyright © 2010 John Wiley & Sons, Inc.

A

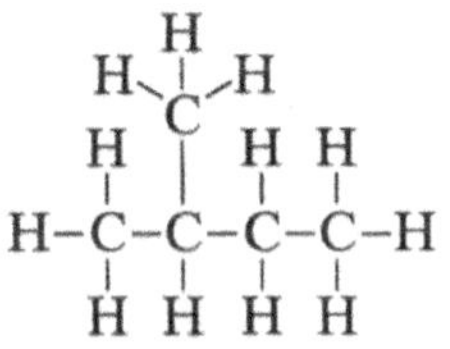

2-methylbutane 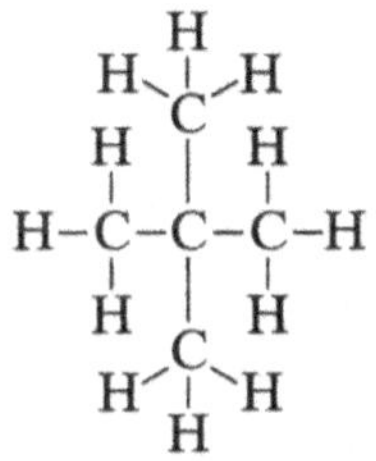   2,2-dimethylpropane

These two compounds are structural isomers of pentane because they have the same molecular formula (C_5H_{12}) but different structures. ∎

Q Do these compounds have the same composition? How do the compounds differ?

A They have the same molecular formula and must therefore have the same percentage of carbon and hydrogen. The compounds differ in their atom connectivities; that is, they differ in their structures. You could also say that they differ in the way the carbons are branched, and they are sometimes called *branched* or *skeletal isomers*. They also differ in their names and chemical and physical properties. ∎

Q Is the following structure another structural isomer?

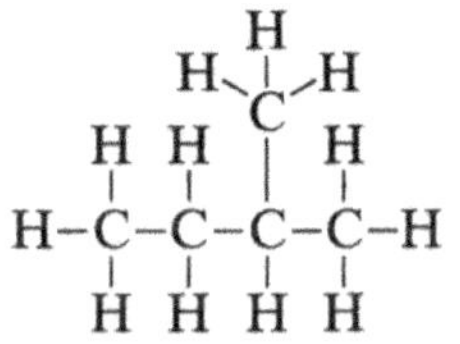

```
        H
     H \ | / H
  H  H   C   H
  |  |   |   |
H-C - C - C - C - H
  |  |   |   |
  H  H   H   H
```

A Careful examination of this structure will reveal that it is the same compound (2-methylbutane) as the second one that we drew. In fact, this one is just the second one flipped 180° horizontally.

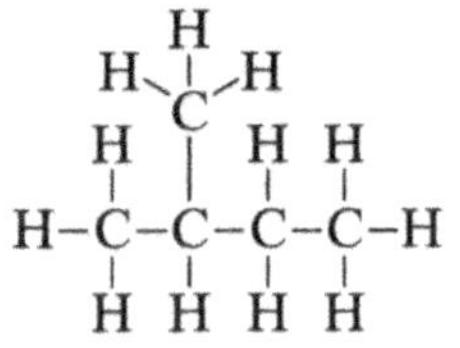

```
        H
     H \ | / H
  H      C      H H
  |      |      | |
H-C  -  C  -  C - C - H
  |      |      | |
  H      H      H H
```

Q Draw structural isomers of hexane (C₆H₁₄), and name them.

A

CH₃CH₂CH₂CH₂CH₂CH₃
hexane

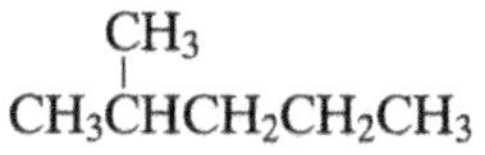

        CH₃
        |
CH₃CHCH₂CH₂CH₃
2-methylpentane

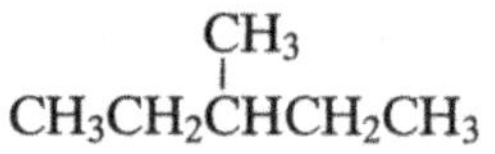

     CH₃
     |
CH₃CH₂CHCH₂CH₃
3-methylpentane

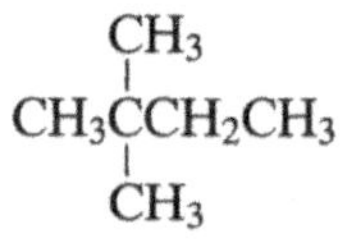

     CH₃
     |
CH₃CCH₂CH₃
     |
     CH₃
2,2-dimethylbutane

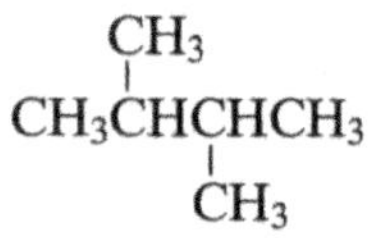

    CH₃
    |
CH₃CHCHCH₃
      |
      CH₃
2,3-dimethylbutane

Q Draw the structure of a branch isomer of 1-chlorobutane.

A

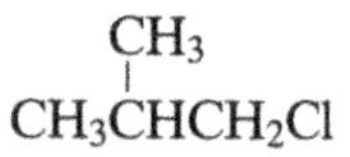

    CH₃
    |
CH₃CHCH₂Cl
1-chloro-2-methylpropane

Q Now, let's tackle butanal, which has a molecular formula of C₄H₈O. Write a structural formula for each of the possible structural isomers of butanal.

$$CH_3CH_2CH_2\overset{\overset{\displaystyle O}{\|}}{C}H$$

A First, we can see that it is possible to maintain the same functional group, but change the carbon skeleton. The compound below, 2-methylpropanal, is a *branch* isomer of the compound because we have moved a carbon in the straight chain to make the chain branched.

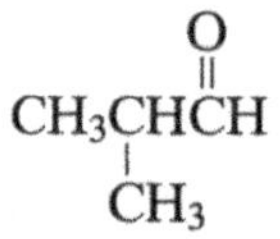

$$CH_3\underset{\underset{\displaystyle CH_3}{|}}{CH}\overset{\overset{\displaystyle O}{\|}}{C}H$$

Q Is there any other way to change the carbon skeleton while maintaining the same functional group?

A It is tempting to change the skeleton by moving one carbon over to the right side of the carbonyl to form the compound

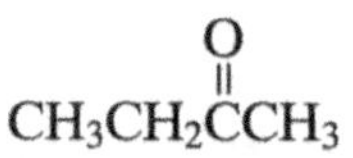

$$CH_3CH_2\overset{\overset{\displaystyle O}{\|}}{C}CH_3$$

but this changes the functional group from an aldehyde (which must have a hydrogen attached to the carbonyl) to a ketone, which always has two carbons attached to the carbonyl. Thus, there is no other structure that we can draw that has the aldehyde functional group *and* the molecular formula C₄H₈O. In order to create other isomers, we must turn to other functional groups, such as the ketone shown above. Because there is only one oxygen in the formula of our original compound, we are restricted to compounds with a functional group that contains only one oxygen, such as aldehydes, ketones, alcohols, and ethers. We have found two aldehyde structural isomers and one ketone isomer and must now turn to alcohols.

Q Write a structural formula for a structural isomer of butanal that is an alcohol.

A An obvious formula to try is the one shown below containing an OH group attached to a carbon. Because each carbon must have sufficient hydrogen atoms to satisfy its octet, the molecular formula turns out to be $C_4H_{10}O$.

$$CH_3CH_2CH_2CH_2OH$$

This molecular formula has two more hydrogen atoms than C_4H_8O, the formula of our set of structural isomers, and therefore this compound is *not* one of these isomers. The fact that simple replacement of the aldehyde group with the alcohol group does not produce a structural isomer is due to the fact that the C=O group has a double bond and therefore two fewer hydrogen atoms than the CH–O–H group. However, we could produce a structural isomer using an OH group *if* we remove two hydrogen atoms from some other place in the molecule. For example, we can turn the compound above into a structural isomer by changing a C–C bond into a C=C bond (which requires the removal of two hydrogen atoms).

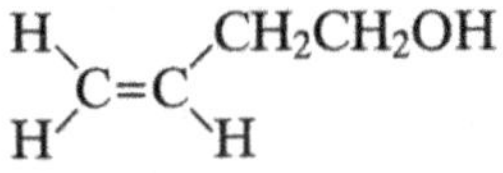

$$\underset{H}{\overset{H}{>}}C=C\underset{H}{\overset{CH_2CH_2OH}{<}}$$

■

Q Use this same strategy to produce another, similar structural isomer.

A We can place the double bond at another point in the molecule, as follows

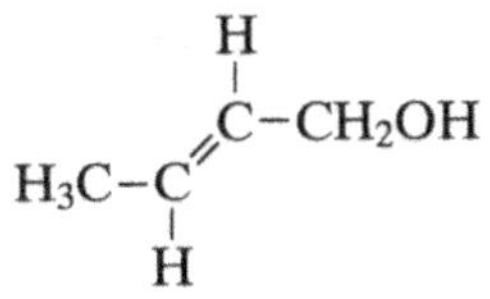

$$H_3C-\underset{H}{\overset{H}{C}}=\underset{H}{\overset{}{C}}-CH_2OH$$

or move the OH group to other positions on the carbon chain.

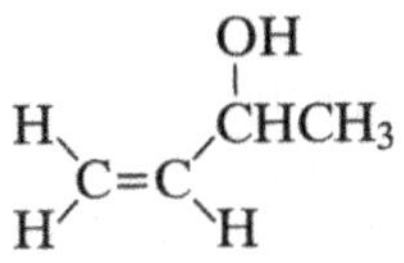

$$\underset{H}{\overset{H}{>}}C=C\underset{H}{\overset{CHCH_3}{<}}\;\;\overset{OH}{|}$$

■

Another method of creating a molecule with two fewer hydrogens is to tie the ends of the carbon chain together in a ring or cycle. Consider the difference between the alkane butane

$$CH_3CH_2CH_2CH_3$$
butane

and its cyclic analog

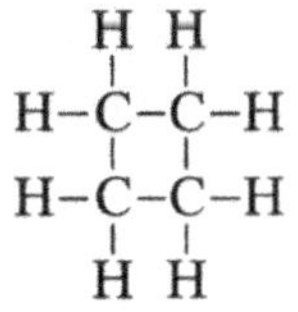

$$\begin{array}{c} H\;\;H \\ |\;\;\;| \\ H-C-C-H \\ |\;\;\;| \\ H-C-C-H \\ |\;\;\;| \\ H\;\;H \end{array}$$
cyclobutane

The cyclic alkane contains two fewer hydrogen atoms than does its acyclic analog. (The molecular formulas are C_4H_{10} for butane and C_4H_8 for cyclobutane.) The formation of the new carbon–carbon bond necessarily results in two fewer hydrogen atoms.

Q Create a structural isomer (with the molecular formula C_4H_8O) that contains the alcohol functional group and a ring.

A

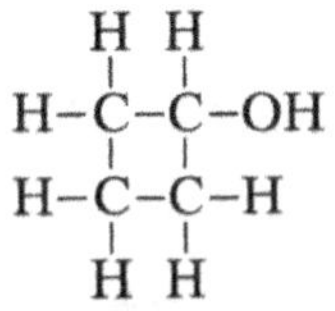

$$\begin{array}{c} H\;\;H \\ |\;\;\;| \\ H-C-C-OH \\ |\;\;\;| \\ H-C-C-H \\ |\;\;\;| \\ H\;\;H \end{array}$$
cyclobutanol

Several other ring isomers, all of which are alcohols with the same molecular formula, are shown below.

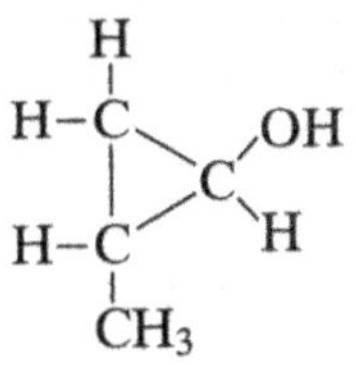

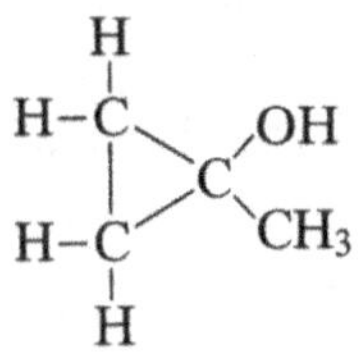

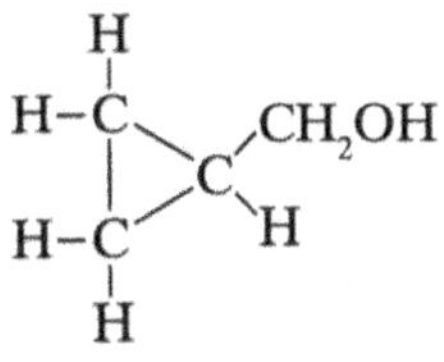

Finally, we consider the ether functional group, which must contain a carbon on both sides of the oxygen (note that the ketone is to the aldehyde as the ether is to the alcohol). Some of the possible structural isomers containing the ether functional group are shown below.

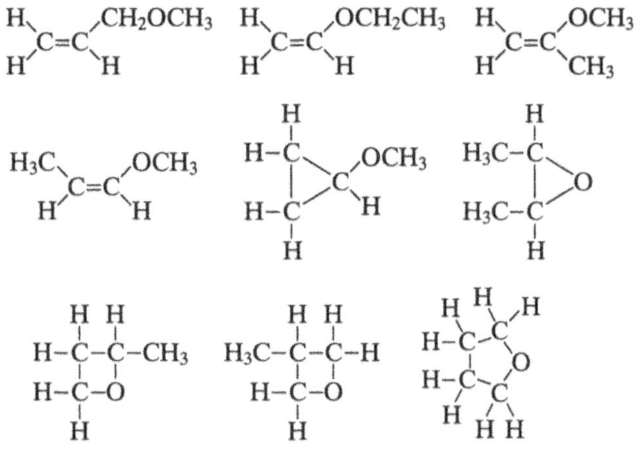

54 STRUCTURE, ISOMERISM, AND STEREOCHEMISTRY

$$\underset{H}{\overset{H}{>}}C=C\underset{H}{\overset{CH_2OCH_3}{<}} \quad \underset{H}{\overset{H}{>}}C=C\underset{H}{\overset{OCH_2CH_3}{<}} \quad \underset{H}{\overset{H}{>}}C=C\underset{CH_3}{\overset{OCH_3}{<}}$$

$$\underset{H}{\overset{H_3C}{>}}C=C\underset{H}{\overset{OCH_3}{<}} \quad \begin{array}{c} H \\ H-C \\ | \\ H-C \\ | \\ H \end{array}\!\!\!\!\!C\!\!\!\begin{array}{c} OCH_3 \\ \\ H \end{array} \quad \begin{array}{c} H \\ H_3C-C \\ | \\ H_3C-C \\ | \\ H \end{array}\!\!\!\!\!O$$

$$\begin{array}{cc} H\ H \\ |\ | \\ H-C-C-CH_3 \\ |\ \\ H-C-O \\ | \\ H \end{array} \quad \begin{array}{cc} H\ H \\ |\ | \\ H_3C-C-C-H \\ |\ \\ H-C-O \\ | \\ H \end{array} \quad \begin{array}{c} H\ \ H \\ H-C\!\!-\!\!C\!\!\diagup\!\!H \\ |\ \ \ \ |\ \ \ \ O \\ H-C\!\!-\!\!C\!\! \\ |\ \ \ \ |\ \\ H\ \ H \end{array}$$

We have now generated 19 structural isomers of the original aldehyde. This set of 20 compounds all have the molecular formula C_4H_8O.

Q How many structural isomers of ethanol are there?

A With only two carbons in the carbon chain, there is no other way to arrange this carbon skeleton. There is also no way to create a new alcohol; specifically, moving the OH to the "other" carbon atom merely produces the same molecule (turned 180° in the plane of the paper). We are forced therefore to change the functional group. If we were to use the oxygen-containing functional groups that contain the carbonyl group, our molecule would not have enough hydrogen atoms. Ethanal is therefore not a structural isomer of ethanol.

$$\underset{CH_3CH}{\overset{O}{\overset{\|}{}}}$$

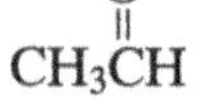

However, the ether functional group does not contain any double bonds and should produce a compound containing two carbons with the correct number of hydrogen atoms. The compound CH_3OCH_3, dimethyl ether, is the one and only structural isomer of ethanol. ∎

Q Write a structural isomer of acetic acid that has the aldehyde functional group.

A Acetic acid contains only two carbons and there is therefore only one structural isomer that is an aldehyde. The isomer can be generated by moving the OH attached to the carbonyl carbon in acetic acid to the sp^3 carbon, thereby producing 2-hydroxyethanal.

$$\underset{HOCH_2CH}{\overset{O}{\overset{\|}{}}}$$

∎

Let's look at a practical application of this process of determining structural isomers. Tanning ointments contain a compound that has the molecular formula $C_3H_6O_3$. In organic chemistry you will learn to use infrared spectroscopy to determine that this compound contains two functional groups—an alcohol and a ketone. We begin our attempt to write a structural formula for this compound by using the ketone functional group. Because there are only three carbons and a ketone must contain a carbonyl *and* at least one carbon on each side of the carbonyl, we can be certain that the backbone of the compound looks like this:

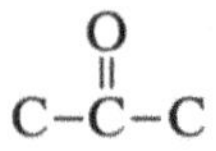

$$\underset{C-C-C}{\overset{O}{\overset{\|}{}}}$$

Now, we just need to attach at least one OH. Because an OH group is always attached to a carbon with a single bond, we could write the following:

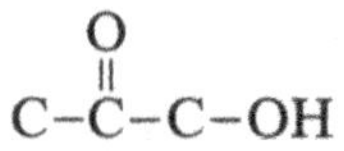

$$\underset{C-C-C-OH}{\overset{O}{\overset{\|}{}}}$$

However, the molecular formula requires a total of three oxygen atoms, so apparently we must use two OH groups. Consequently, there are two possibilities for the compound, the symmetrically substituted compound

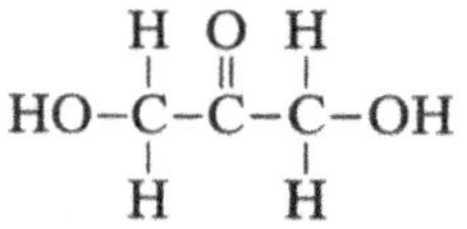

$$\begin{array}{c} H\ \ O\ \ H \\ |\ \ \ \|\ \ \ | \\ HO-C-C-C-OH \\ |\ \ \ \ \ \ \ | \\ H\ \ \ \ \ H \end{array}$$

or the unsymmetric compound

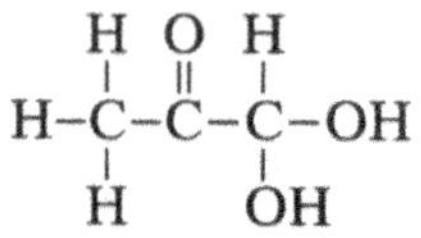

$$\begin{array}{c} H\ \ O\ \ H \\ |\ \ \ \|\ \ \ | \\ H-C-C-C-OH \\ |\ \ \ \ \ \ \ | \\ H\ \ \ \ OH \end{array}$$

In fact, the active ingredient in some tanning oils is 1,3-dihydroxypropanone, the symmetric isomer shown above. This step-by-step procedure is exactly what chemists use to obtain the structure of a compound from the various bits of information, such as the type of functional group and the number of carbons that can be obtained from a variety of chemical and physical (particularly spectroscopic) techniques.

Q Draw and name the three structural isomeric ketones of pentanal.

$$CH_3CH_2CH_2CH_2\overset{\overset{O}{\|}}{C}H$$

A Two are straight-chain ketones; one is a branched ketone.

$$CH_3CH_2CH_2\overset{\overset{O}{\|}}{C}CH_3 \qquad CH_3CH_2\overset{\overset{O}{\|}}{C}CH_2CH_3 \qquad CH_3\overset{\overset{O}{\|}}{C}\underset{\underset{CH_3}{|}}{C}HCH_3$$

2-pentanone   3-pentanone   3-methyl-2-butanone ∎

Q A compound has the formula $C_3H_4O_2$ and is a carboxylic acid. Write a structural formula for it.

A

∎

STEREOISOMERISM

Although structural (or skeletal) isomers are easily identified by the presence of different carbon skeletons or different functional groups, stereoisomers have the same atom-to-atom sequence (or connectivity) but different spatial arrangements of the atoms in three dimensions. There are two different types of stereoisomers: *geometric* isomers and *optical* isomers.

Geometric Isomers

Geometric isomers have easily recognizable three-dimensional differences. Consider the alkene 2-butene. Because the C=C bond is rigid (see the earlier discussions on bonding at the C=C bond on page 47), the four carbon atoms all lie within the same plane in fixed positions. Therefore, two arrangements of the atoms in the molecule are possible. The carbons attached to the C=C linkage can reside on the same side of the molecule or can be opposite one another. These two arrangements of the atoms cannot be converted into one another under ordinary conditions, and therefore two different compounds with the same connectivities exist. The compound with the methyl groups on the same side is called the *cis* isomer, and the other is the *trans* isomer (refer to the section titled Alkene Geometric Isomers in Chapter 2). Geometric isomers have different physical properties (e.g., the boiling point of *cis*-2-butene is 3.7 °C, while the boiling point of *trans*-2-butene is 0.9 °C).

trans-2-butene   cis-2-butene

Q The boiling points of the 2-butene geometric isomers are very close. Can you rationalize this observation?

A To answer this question, you must review the forces between molecules. There are three types of intermolecular forces: (1) van der Waals (also called *London dispersion*) forces, (2) dipole–dipole forces, and (3) hydrogen bonding. *Van der Waals forces* are due to the instantaneous dipole-induced dipole interaction that occurs between all molecules (as well as atoms and ions). *Dipole–dipole forces* exist between two molecules that have dipole moments due to the alignment of the positive end of the dipole of one molecule with the negative end of another molecule. *Hydrogen bonding*, the strongest intermolecular force, occurs between molecules that have a hydrogen attached to an electronegative atom such as oxygen, nitrogen, and fluorine.

In the case of hydrocarbons, the molecules generally have very low dipole moments and no hydrogen bonding. Hence, the 2-butene isomers have only van der Waals forces. These forces depend primarily on the number of electrons in a molecule and secondarily on the distance between the molecules. Because the geometric isomers have the same number of electrons, the attraction between two *trans*-2-butene molecules is approximately the same as the attraction between two *cis*-2-butene molecules at a particular separation. Consequently, the amount of energy required to separate the molecules of a liquid and allow the molecules to enter the gaseous state is about the same for both isomers. ∎

Q Determine whether 1,1-dichloro-1-propene can exist as geometric isomers.

A We first need to write the following structural formula for this compound.

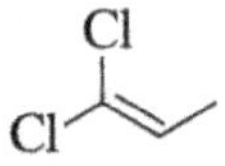

Because of the presence of the C=C bond, we suspect that the groups attached to the C=C linkage can be arranged differently. In the diagram below, we have moved the methyl group from one side of the C=C to the other side (of course, the hydrogen atom on carbon 2 (C2) must move as well).

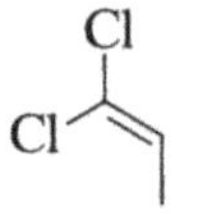

The two structures produced by this three-dimensional movement of the groups represent the same molecule. In order to demonstrate to yourself that the two molecules are superimposable, rotate the image 180° around the line running through the C=C bond. Thus, 1,1-dichloro-1-propene does not exhibit geometric isomerism.

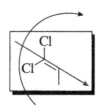

The presence of a C=C double bond is not a sufficient condition for the existence of geometric isomers. In order to exhibit geometric isomerism, the *two groups attached to each carbon of the C=C linkage must be different*. For example, on one of the carbon atoms, the two groups might be H and Cl, and on the other carbon of the C=C linkage the two groups might be CH_3 and C_2H_5, as shown below.

Q Does 1-butene (but-1-ene) or 2-pentene (pent-2-ene) exhibit geometric isomerism?

A But-1-ene has the same atom (H) attached to one of the carbons of the C=C linkage. Pent-2-ene has a methyl group and a hydrogen attached to one carbon of the C=C linkage and an ethyl group and a hydrogen attached to the other carbon. Therefore, only pent-2-ene exhibits geometric isomerism. ∎

Another type of "rigidity" in a molecule can produce geometric isomers. A ring of carbons prevents the molecule from converting cis or trans isomers into one another. Consider 1,2-dimethylcyclobutane:

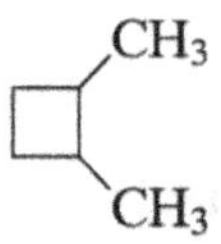

The two methyl groups can be on either the same side (*cis*) or opposite sides (*trans*) of the ring. (Remember that a solid wedge indicates a bond coming toward the viewer and a dashed wedge indicates a bond pointing away from the viewer.)

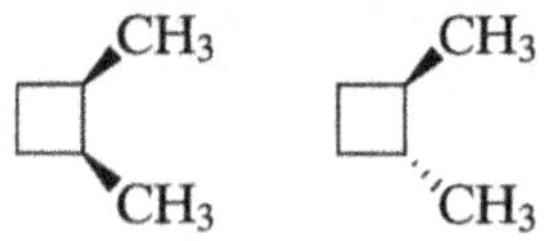

The side view of the ring may make the stereochemistry more obvious:

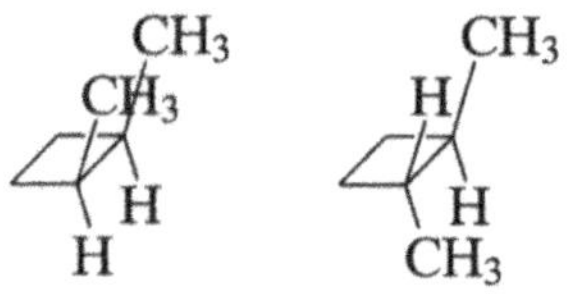

cis   trans

Q Determine which of the following compounds exhibit geometric isomerism: 1-chloro-2-methylpropene, 1-bromo-1-pentene, or 1,2-difluorocyclopropane.

A Two of these compounds can exhibit geometric isomerism: 1-bromo-1-pentene and 1,2-difluorocyclopropane. Because 1-chloro-2-methylpropene has two identical groups on one of the carbons of the C=C linkage, it cannot exhibit geometric isomerism.

∎

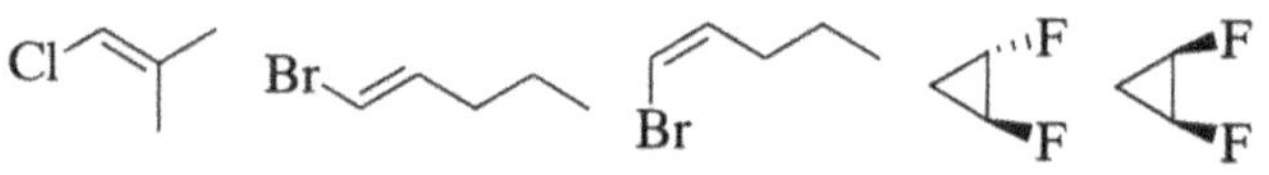

Q Try your hand at drawing a side view of the ring of *cis*-1,2-difluorocyclopropane.

A

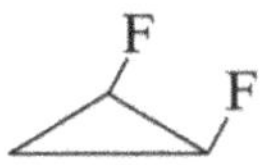

Q Does 1,1-dimethylcyclobutane exhibit geometric isomerism?

A No, the interchange of the methyl groups simply produces the same molecule. Note that the movement of methyl groups, for example, to produce 1,2-dimethylcyclobutane, changes the atom connectivities (both methyl groups are no longer attached to the *same* carbon). This change therefore produces a structural isomer, not a geometric isomer of 1,1-dimethylcyclobutane. ∎

Optical Isomerism

The second type of stereoisomerism, *optical isomerism*, has a more subtle 3D relationship among the atoms than is the case in geometric isomerism. The sole criterion for this type of three-dimensional isomerism is as follows.

Statement. Optical isomers occur when the mirror image of one molecule is not superimposable on that molecule.

The two nonsuperimposable mirror images, as we will see below, differ in their effect on the rotation of plane-polarized light. These non-superimposable mirror images are called *enantiomers* or *optical isomers*. They often arise when a molecule contains a carbon with four different groups attached to it. Such a carbon is said to be *chiral* (from the Greek *chiros*, for hand). The enantiomers of 1-chloro-1-methoxyethane are shown below.

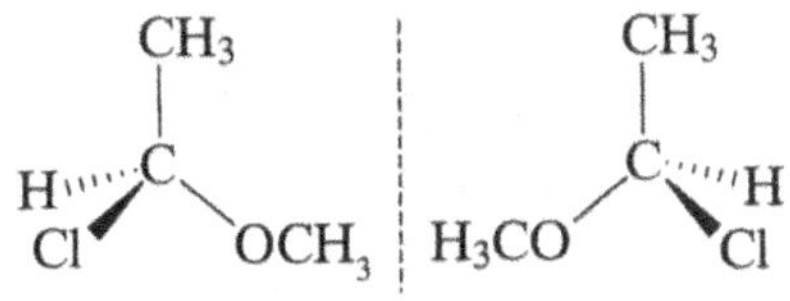

Nonsuperimposable mirror images (enantiomers) of 1-chloro-1-methoxyethane. The vertical line represents a mirror plane perpendicular to the plane of the paper.

Optical isomers sometimes are said to exhibit the property of "handedness" by analogy to the relationship between left and right hands. If you hold the palm of your right hand toward a mirror, the mirror image will be identical to a view of the palm of your left hand. Thus, your hands are mirror images of one another. If you try to superimpose them with the palms facing in the same direction, this cannot be accomplished—your thumbs will be pointing in opposite directions. They are nonsuperimposable mirror images and are therefore enantiomeric in their relationship to each other in three-dimensional space.

Q Which of the following exhibit optical isomerism? Carefully draw and examine the mirror images. It may be helpful to use molecular models.

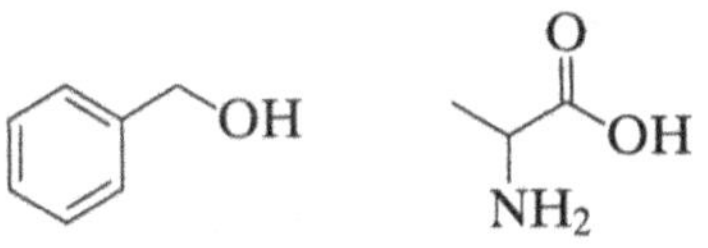

benzyl alcohol   alanine

A Benzyl alcohol has no carbon that contains four different groups attached to it. In alanine the α carbon has four different groups: the COOH group, the CH_3 group, the NH_2 group, and hydrogen. For benzyl alcohol the mirror images are superimposable; that is, the two images represent identical molecules. For the amino acid alanine the mirror images cannot be superimposed (make the models to prove to yourself that no manner of turning or rotating one of the images will make it superimposable on the other).

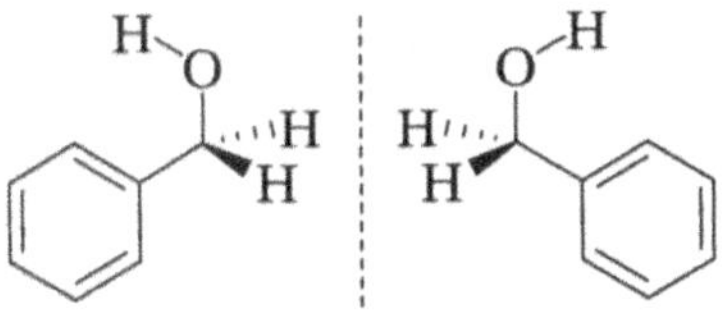

benzyl alcohol

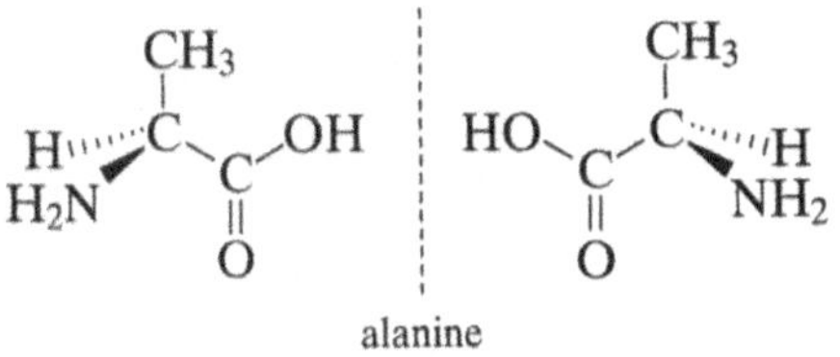

alanine ∎

The relationship between two molecules can also be determined by reference to an isomeric tree diagram.

58 STRUCTURE, ISOMERISM, AND STEREOCHEMISTRY

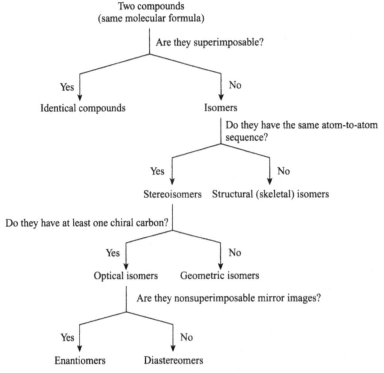

Figure 4.1. Classification of isomers.

The flowchart diagram in Figure 4.1 presents questions, the answers to which lead sequentially to the classification of either structural or stereoisomers. The optical isomers appear at the bottom of the tree, where a new word appears—*diastereomer*.

Statement. A pair of isomeric molecules are diastereomers of each other if they have two or more chiral carbons at the same positions but are not enantiomers; that is, diastereomers are stereoisomers that are not mirror images.

Q An α-amino acid is a compound in which both a carboxylic acid and an amino functionality are attached to the same atom. Glycine ($H_2NCH_2CO_2H$) is the simplest α-amino acid. Apply the tree diagram to the mirror images of the amino acid 2-aminopropanoic acid, more commonly called alanine.

A The mirror images of alanine are shown again below:

According to the tree diagram, we must first determine whether these two molecules are superimposable. A simple way to do this is to make models of the two molecules and then try to get each part of one molecule to "sit" on top of the analogous part of the other molecule. Below we show an attempt to do this, and it should be clear that at least this maneuver does not allow superimposition of the mirror images.

Moreover, it is true that all such attempts fail to make the two molecules superimposable. The molecules are therefore isomers. Next, we determine whether they have the same atom connectivity. Because the nitrogen is connected to the carbon adjacent to the carbonyl group in both molecules and the same comparisons can be made for every other group, we conclude that the two molecules are not structural isomers but are instead stereoisomers. Next, we search for a chiral carbon, that is, a carbon that has four different groups around it—the carbon labeled with an asterisk below.

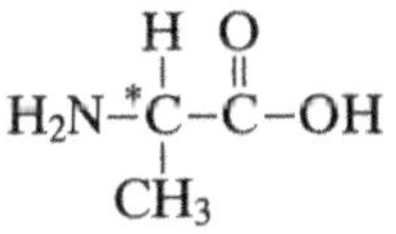

$$H_2N-\overset{*}{\underset{CH_3}{\overset{H}{C}}}-\overset{O}{\overset{\|}{C}}-OH$$

(where * indicates a chiral carbon; i.e., one with four different groups). The molecules therefore must be optical and not geometric isomers. Finally, we must apply the criterion of nonsuperimposability to these mirror images (again). As we know, they are not superimposable and therefore are enantiomers. ∎

Before we work on an example that exhibits diastereoisomerism, we need to know that optical isomers can be distinguished by their effects on plane-polarized light. Although monochromatic light (light of one wavelength) normally consists of electromagnetic waves radiating in all directions, a polarizing lens transmits light only in a single plane. When this light passes through a solution that contains just one enantiomer of a compound, the plane of the light is rotated either clockwise or counterclockwise. This effect can be observed in a device called a *polarimeter*. If the other enantiomer is placed in the polarimeter, the plane-polarized light rotates in the opposite direction by the same number of degrees. The direction of rotation is used to designate one stereoisomer as *levorotatory* if the light rotates in a counterclockwise direction and as *dextrorotatory* if it rotates in a clockwise direction. Hence, the enantiomer of alanine that rotates plane-polarized light in the counterclockwise direction is referred to as *l*-alanine, and the isomer that rotates plane-polarized light in a clockwise direction is *d*-alanine.

Q How would an equal mixture of enantiomers affect plane-polarized light?

A Plane-polarized light is not affected by this mixture, which is called a *racemic mixture*. In the mixture the *l*-enantiomer rotates the plane of the polarized light by the same number of degrees in the counterclockwise direction as the *d*-enantiomer rotates the light in the clockwise direction. The result is that the plane of the light appears not to have moved. ∎

Louis Pasteur's (French microbiologist and chemist, 1822–1895) work on salts of tartaric acid, byproducts of the wine industry in France, led to the isolation of the stereoisomers of tartaric acid. Pasteur noticed that the surface features of some crystals appeared to have a mirror-image relationship to that of other crystals. The crystals were sufficiently large that he was able to separate them using a tweezers. After separation, he found that a solution of one set of crystals rotated plane-polarized light clockwise (dextrorotatory) whereas the solution produced from the crystals of the mirror image rotated plane-polarized light counterclockwise (levorotatory). This serendipitous discovery of a macroscopic crystalline property that is a consequence of a molecular property is a rare occurrence. Pasteur referred to the stereoisomers as *dissymmetric*, but the British physicist Lord Kelvin (William Thomson, 1824–1907) coined the term *chiral*. Pasteur subsequently determined that combining equivalent quantities of the dextrorotatory and levorotatory tartaric acid produced a sample that did not rotate plane-polarized light. Racemic acid, an equal mixture of the two enantiomers of tartaric acid, is designated by the symbol (±) or (*d/l*).

Tartaric acid is 2,3-dihydroxybutanedioic acid

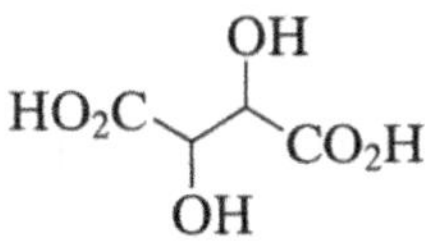

$$HO_2C\underset{OH}{\overset{OH}{\diagup\!\!\!\diagdown}}CO_2H$$

and has two chiral centers, that is, two carbons that have four different groups surrounding them. The molecule is redrawn below in 3D perspective:

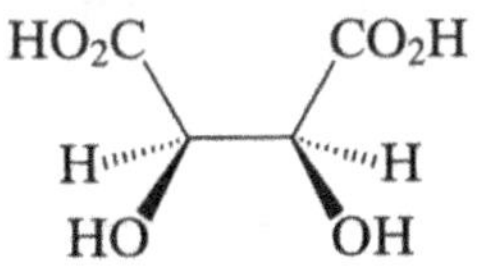

$$\begin{array}{c} HO_2C \qquad\qquad CO_2H \\ H^{\text{\tiny{\textbackslash\textbackslash\textbackslash}}}\qquad\qquad{}^{\text{\tiny{///}}}H \\ HO \qquad\qquad OH \end{array}$$

Q Does this molecule as drawn have a nonsuperimposable mirror image?

A For most students the use of a molecular model is necessary to answer this question, but the diagram below should help you to draw the mirror image.

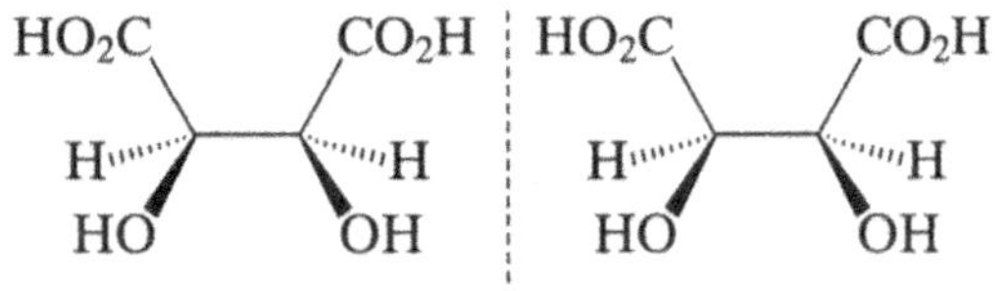

60 STRUCTURE, ISOMERISM, AND STEREOCHEMISTRY

You can determine that the molecule on the right of the mirror is superimposable on the one on the left by simply translating (moving) it from right to left, directly across the mirror. Thus, the mirror images are not enantiomers but are instead identical. If you refer to the tree diagram above you will find that a molecule that has two or more chiral centers and has a superimposable mirror image is a diastereomer of other optical isomers. This compound is called a *meso compound*; it is *meso*-tartaric acid.

Statement. A *meso* compound is a compound with two or more chiral centers for which its mirror image is itself.

We are now going to change the *configuration* (the 3D arrangement) of the groups at the left-hand carbon. The order of the groups can be changed by simply interchanging two groups, as, for example, in the schematic example below:

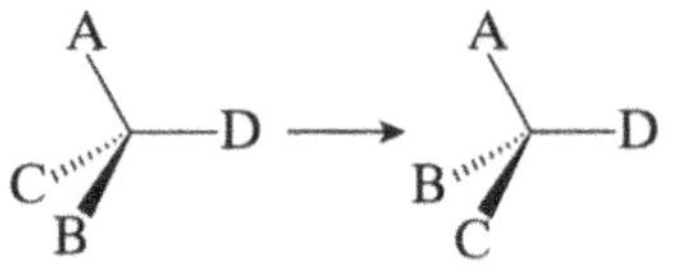

Exchange of the H and OH at the left-hand carbon in *meso*-tartaric acid produces the following compound:

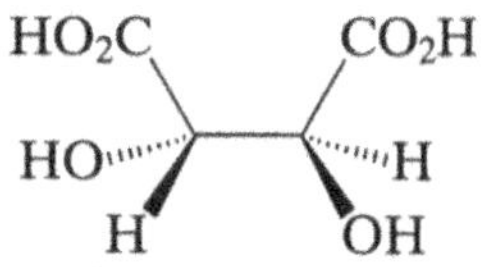

Q Draw the mirror image of this molecule and determine whether it has a nonsuperimposable mirror image.

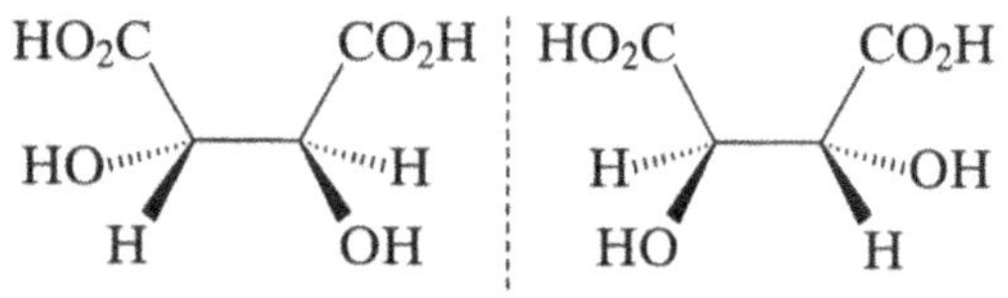

Note that the change in the order of the groups at one carbon produces a nonsuperimposable mirror image. These two molecules are therefore enantiomers. ■

Hence, tartaric acid has more than two stereoisomers that result from the presence of two chiral carbons. The relationship between the *maximum* possible number of stereoisomers and the number of chiral carbons n is 2^n. After a brief interlude on the naming of enantiomers, we will return to a method used to represent enantiomers and diastereomers.

Absolute Configuration

Enantiomers can be designated by the direction in which polarized light is rotated, but this would require an experimental procedure every time we want to name a compound that has an enantiomer. To avoid the need for an experimental basis for the name, a system of naming enantiomers is based on the Cahn–Ingold–Prelog (CIP) sequence rules.

Statement. The four different substituents on a chiral carbon are prioritized ($a > b > c > d$) based on atomic numbers; higher atomic numbers are given higher priority. The molecule is then oriented spatially so that the chiral carbon is in front of the lowest-priority **d** substituent. By tracing an arc from **a** to **b** to **c**, one can assign as R the enantiomer for which the rotation is clockwise and as S the enantiomer for which the rotation is counterclockwise. The designations R and S come from the Latin for *rectus* (right) and *sinister* (left).

Q Let's consider the simple molecule

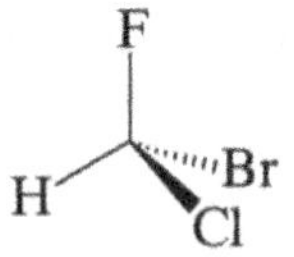

which clearly has four different substituents, thereby making the carbon chiral. Assign this molecule as either R or S.

A The atomic numbers vary in the order Br > Cl > F > H, and therefore the lowest-priority substituent is H. The molecule shown above must now be oriented so that the carbon is in front of the hydrogen. The following structure shows the molecule rotated so that the hydrogen is behind the carbon:

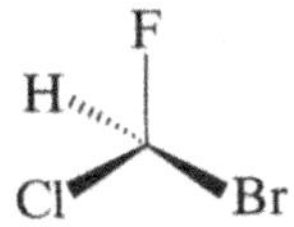

Drawing an arc from bromine (highest priority) to chlorine to fluorine produces a clockwise arc, and this enantiomer is therefore designated as the *R* enantiomer. ∎

Now let's return to alanine and assign priorities to the four different substituents in alanine.

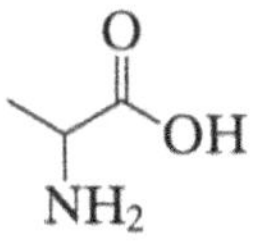

Because nitrogen has a higher atomic number than either carbon or hydrogen, the amino group has the highest priority, **a**. Both the carboxylic acid and methyl substituents, however, are attached to the chiral carbon through a carbon. In such a case we must look at the atoms attached to each of these carbons to assign priorities. The carboxylic acid, for these purposes, is equivalent to a carbon surrounded by three oxygen atoms due to the C=O (a carbon–oxygen double bond is counted as two oxygens) and C–O bonds.

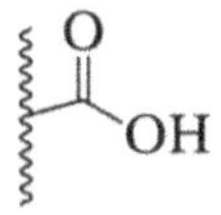

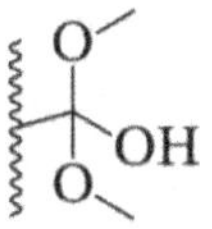

The methyl group contains a carbon with three attached hydrogens. Because the atomic number of oxygen is higher than that of hydrogen, the carboxylic acid has a higher priority than does methyl; that is, COOH has priority **b** and CH₃ is **c**. Hydrogen has the lowest priority **d**. Returning now to the two enantiomers of alanine, we can see that the enantiomer on the left is (*R*)-2-aminopropanoic acid; the one on the right is (*S*)-2-aminopropanoic acid

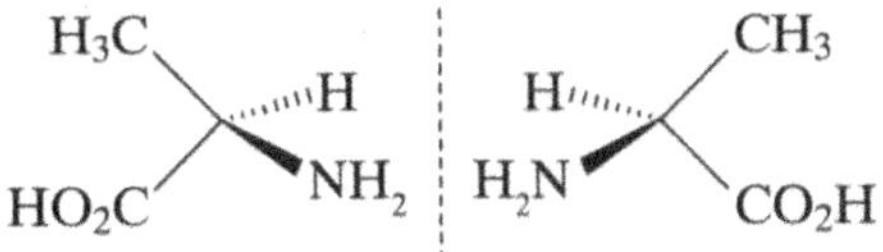

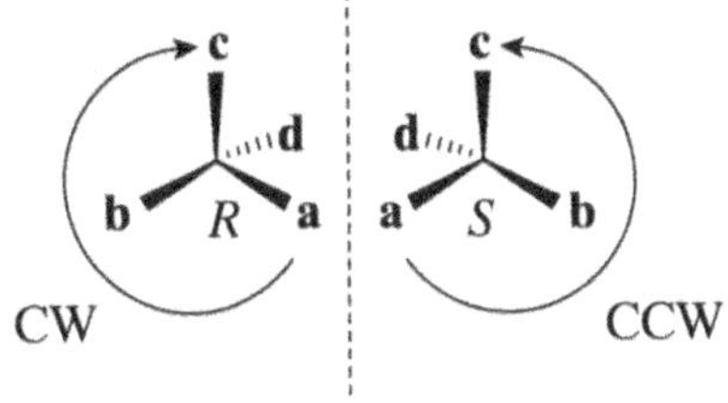

(where CW = clockwise; CCW = counterclockwise).

Q Provide the name of this specific stereoisomer of 2-chlorobutane (*sec*-butyl chloride) shown below:

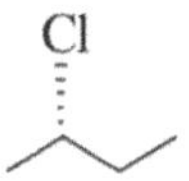

A The molecule is drawn as a line notation structure with the four carbon atoms in the plane of the paper and the chlorine substituent pointing behind the plane, as indicated by the dashed line. The hydrogen attached to the chiral carbon must be coming toward us out of the plane of the paper.

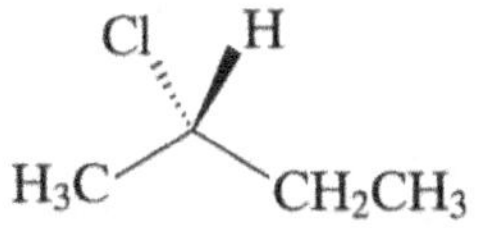

The CIP priorities are as follows: Cl > CH₂CH₃ > CH₃ > H. (The ethyl group has a higher priority than does methyl because the carbon attached to the CH₂ in the ethyl group has a higher atomic number than does hydrogen.) Hydrogen, with the lowest priority, is in front instead of in back of the chiral carbon. The simplest way to deal with this difference is to recognize that the direction of rotation obtained by the **a → b → c** rotation will be reversed if the molecule is turned around in order to move the hydrogen behind the carbon. Thus, the clockwise **a → b → c** rotation produced by moving from the highest priority Cl to the CH₃ (with the molecule as drawn) must be reversed to counterclockwise, and the compound is (*S*)-2-chlorobutane. Alternatively, the molecule can be redrawn so that hydrogen is behind carbon, from which the *S* designation is clear.

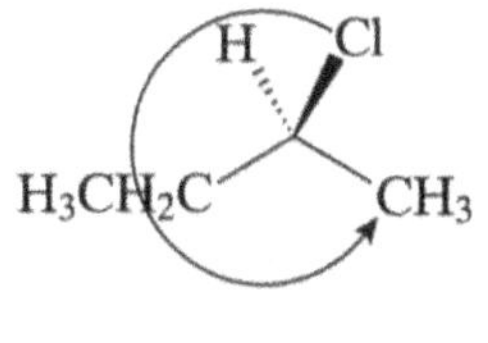

∎

Q Assign the absolute configuration to the chiral carbon in each structure.

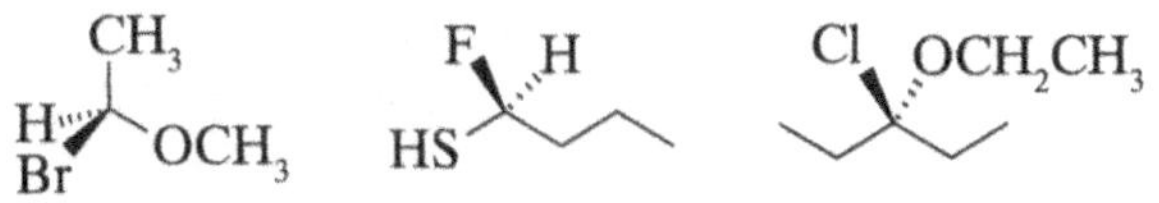

(a)   (b)   (c)

A In **a** the absolute configuration is *S*. The absolute configuration in **b** is *R*. Compound **c** has no chiral carbon. ∎

Fischer Projections

There is a simpler way to represent two enantiomers without drawing wedges and dashed lines to represent the two bonds in front of and behind the plane of the paper, respectively. The Fischer projection was devised by the German chemist Emil Fischer (1852–1919), who received the Nobel Prize in Chemistry in 1902. In a Fischer projection a chiral carbon is shown as the intersection of vertical and horizontal line segments.

Statement. In a Fischer projection the horizontal substituents are assumed to point forward (as shown by the wedges in the equivalent drawing), and the two vertical substituents are assumed to point back (as shown by the dashed lines in the equivalent drawing).

The advantage of the Fischer projection is the ease of drawing an enantiomer. The disadvantage is that it is a 2D drawing of a 3D molecule. Thus, caution must be exercised when we determine which is the *R* enantiomer and which is the *S* enantiomer. If we place the H (priority **d**) in a vertical position, this substituent is behind the chiral carbon, as it should be. In this case, the **a** → **b** → **c** rotation will yield the correct *R* or *S* configurational assignment.

$$+\ \ \text{is equivalent to}\ \ +$$

If we place the H at a horizontal position in the Fischer projection, it is pointing forward. For the enantiomer of alanine on the right below, the clockwise direction of the **a** → **b** → **c** rotation must be reversed because an observer looking at the molecule from the correct perspective with the H behind the chiral carbon would view the **a** → **b** → **c** rotation as counterclockwise. Hence, this is the *S* (not the *R*) enantiomer.

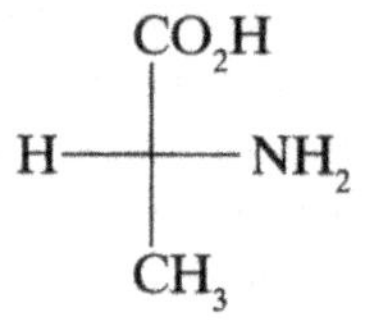

$$\begin{array}{cc}
\text{CO}_2\text{H} & \text{CO}_2\text{H} \\
\text{H}-\!\!\!\!\!\!+\!\!\!\!\!-\text{NH}_2 & \text{H}_2\text{N}-\!\!\!\!\!\!+\!\!\!\!\!-\text{H} \\
\text{CH}_3 & \text{CH}_3 \\
(R)\text{-alanine} & (S)\text{-alanine} \\
(R)\text{-2-aminopropanoic acid} & (S)\text{-2-aminopropanoic acid}
\end{array}$$

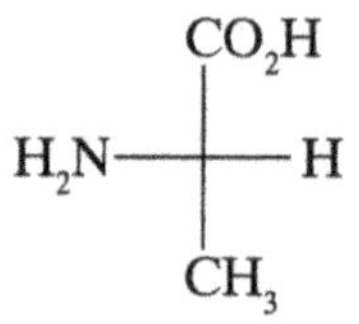

5

CHEMICAL REACTIVITY

Thus far we have discussed primarily the structure of molecules and have not dealt with their reactivity. The tendency of a particular molecule to react with some reagent depends on the functional groups, the type of bonding, and the substituents. Because most reactions are performed in a solvent under a certain set of temperature and pressure conditions, the nature of the reaction will also depend on these variables.

RATE VERSUS EXTENT OF REACTION

The reaction of a molecule with a given reagent is measured by two very important factors: the rate of the reaction and the extent of the reaction.

Statement. The *rate* of a reaction is a measure of the amount of reactant consumed in a certain time (and the amount of products yielded by the reaction in a certain time). The *extent* of the reaction is a measure of how much reactant has been converted to products by the time equilibrium has been reached.

Let's consider a hypothetical example of molecule A in a reaction with molecule B to form molecules C and D:

$$A + B \rightleftharpoons C + D$$

This reaction is characterized by the experimental data shown in Table 5.1 with the concentrations of reactants and products shown at different times.

Note that as the reaction time approaches 5 h, the concentration of A begins to level off and approach the equilibrium value of 0.05 M. The reaction above is an example of a reaction that reaches equilibrium at a modest rate (fast reactions occur in milliseconds or less) and has a high extent (most of the reactants are converted to products).

Some reactions have such small extents or such low rates that they appear to never reach equilibrium. For example, the decomposition of benzene to elemental carbon and hydrogen (H_2) is very favorable, and 99.999% of benzene will eventually decompose to its constituent elements. However, this reaction is so slow that it would take many years for even 10^{-10} mol of benzene to revert to carbon and hydrogen. Thus, this reaction appears to be a nonequilibrium reaction.

Some reactions can be *forced* to occur under nonequilibrium conditions. For example, the preparation of elemental silicon used to make computer chips is done by mixing sand (SiO_2) with carbon:

$$SiO_2(s) + C(s) \rightarrow Si(s) + CO_2(g)$$

The products of the reaction are solid silicon and carbon dioxide gas. At room temperature this reaction has a very low extent—essentially no elemental silicon is formed if sand is mixed with carbon at room temperature. However, at 1500 °C the reaction will produce a significant amount of elemental silicon. This high temperature not only requires the expenditure of large amounts of energy but also produces corrosion and

---

The Bridge to Organic Chemistry: Concepts and Nomenclature
By Claude H. Yoder, Phyllis A. Leber, and Marcus W. Thomsen
Copyright © 2010 John Wiley & Sons, Inc.

TABLE 5.1. Molar Concentration (M) of Reactants and Products versus Time for A + B → C + D

Time (h)	[A]	[B]	[C]	[D]
0	1.00	1.00	0.00	0.00
1.0	0.22	0.22	0.78	0.78
2.0	0.13	0.13	0.87	0.87
3.0	0.09	0.09	0.91	0.91
4.0	0.06	0.06	0.94	0.94
5.0	0.05	0.05	0.95	0.95
6.0	0.05	0.05	0.95	0.95
7.0	0.05	0.05	0.95	0.95
9.0	0.05	0.05	0.95	0.95

TABLE 5.2. Molar Concentration (M) of Reactants and Products versus Time for $ANO_2 + B \rightarrow CNO_2 + D$

Time (h)	[A]	[B]	[C]	[D]
0	1.00	1.00	0.00	0.00
1.0	0.65	0.65	0.35	0.35
2.0	0.49	0.49	0.51	0.51
3.0	0.39	0.39	0.61	0.61
4.0	0.32	0.32	0.68	0.68
5.0	0.27	0.27	0.73	0.73
6.0	0.24	0.24	0.76	0.76
7.0	0.22	0.22	0.78	0.78
9.0	0.22	0.22	0.78	0.78

deterioration of the reaction chamber. Because the carbon dioxide is a gas, it can easily be removed from the reaction chamber, thereby preventing equilibrium from being established. (Remember that all reactants and products must be present in order for equilibrium to exist.) According to Le Chatelier's principle, removal of the carbon dioxide forces the reaction to produce more elemental silicon in its effort to reestablish equilibrium (which is never allowed to occur), and consequently the reaction can be run at a lower temperature of ~1000 °C. At this lower temperature the extent of the reaction is not very high *if* equilibrium occurs, but is sufficiently high to produce a commercially viable quantity of elemental silicon due to the disruption of the equilibrium process.

The extent of a reaction is expressed by its equilibrium constant, which for the hypothetical reaction shown in Table 5.1 with all reactants and products in solution is 360:

$$K = \frac{[C][D]}{[A][B]} = \frac{[0.95][0.95]}{[0.05][0.05]} = 360$$

Because the products appear in the numerator of the equilibrium expression, larger equilibrium constants are generally associated with greater extents of reaction. It is common practice to express the equilibrium constant for reactions (such as acid–base reactions) as the pK, which is the negative log of the equilibrium constant

$$pK = -\log K$$

Thus, if an acid has an equilibrium constant for its dissociation in water (K_a) of 10^{-10}, then the pK_a is 10. Another acid may have a greater extent of reaction with water and the K_a might be 10^{-3}. The pK_a for this acid is 3. (Note that the smaller pK_a indicates the greater extent of reaction.)

Now let us suppose that we run the reaction above with a compound very similar to A except that a hydrogen in the compound A is replaced by a nitro group. In other words, the nitro group is a substituent (because it substitutes for hydrogen on the parent). We will refer to this compound as ANO_2 and write the reaction:

$$ANO_2 + B \rightleftharpoons CNO_2 + D$$

It is possible that the nitro group, assuming that it is not destroyed in the reaction, could influence the rate or the extent of reaction or perhaps both.

Q Use the data in Table 5.2 to determine whether one or both of these parameters is affected.

A The data indicate that ANO_2 is consumed more slowly (note that we start the reaction with the same number of moles of ANO_2 as those of A in the first reaction) and produces less of the product when equilibrium is established. Hence, the nitro substituent has affected both the rate and extent of reaction. Both the rate and extent are lower for ANO_2 than for A. ∎

In the synthetic laboratory the extent of a reaction under a certain set of conditions is usually measured as the *percent yield*.

Statement. The *yield* is the actual amount (in grams or moles) of the product obtained from a reaction. The percent yield is the amount of product obtained relative to the amount that would have been obtained had the reaction gone to completion (100%).

Suppose that our reaction was run with 10 g of A, a substance for which we will assume a molecular weight of 100. We will also assume that we have excess reagent B, so that we will not have to account for its amount. Let's also assume that product C has a molecular weight of 200. We run the reaction and allow it to proceed until we believe that equilibrium has been established. After isolating the product we find that 10 g of C were obtained. The stoichiometry of the equation tells us that consumption of one mole of A produces one mole of C, and we therefore know that 10 g of A (10 g/100 g/mol = 0.1 mol A) should have produced 0.1 mol of C or 20 g (0.1 mol × 200 g/mol = 20 g C) if the reaction went 100% to completion. This amount of C (20 g) is the theoretical yield. The 10 g yield of C therefore is 50% of the theoretical yield.

Q Assume that you have just run the following reaction:

$$H_2NCONH_2 + H_2O \rightarrow 2\,NH_3 + CO_2$$

Also assume that the reaction of 0.1 mol of urea (H_2NCONH_2) with excess water has produced 0.020 mol of ammonia after 2 h, 0.040 mol after 5 h, 0.050 mol after 8 h, 0.050 mol after 20 h, and 0.050 mol after 48 h. When does the reaction reach equilibrium (assuming that all reactants and products are soluble in water)? What are the yield, the theoretical yield, and the percent yield?

A The reaction reaches equilibrium somewhere between 5 and 8 h or at 8 h. The actual yield is 0.050 mol ammonia. The theoretical yield of ammonia is 0.20 mol (each mole of urea produces 2 mol of ammonia if it is totally consumed). The percent yield is (0.050 mol/0.20 mol) × 100 = 25%. ■

MECHANISM

Statement. The detailed sequence of events at the molecular level that occurs during the formation of products from reactants is called the *mechanism* of a reaction.

Although much of modern organic chemistry involves discussion of mechanism, such detailed sequences of molecular events are probably unknowable. We speculate about mechanism, explore the consequences of using different reactants, and determine the rate of the reaction and the stereochemistry of the products, but after many laborious experiments have been performed we can still only say that a particular pathway or mechanism is *consistent* with the experimental data. We cannot, at least at the present, prove that a reaction occurs by a particular mechanism.

As an illustration of mechanism, consider the reaction of methyl bromide with hydroxide ion. We will discuss this reaction in some detail later, but for now we use it only as an example. As is true for any reaction, we must first ascertain the products of the reaction. In this case, the products are methanol and bromide ion, and we can now write a chemical equation:

$$CH_3Br + OH^- \rightarrow CH_3OH + Br^-$$

By looking at the nature of the reactants and products we can probably assume that the detailed process by which the hydroxide and methyl bromide interact does not involve breaking the C–H bonds. Because the methyl group appears in both the reactant and product, we can be fairly certain that these C–H bonds are not intimately involved in whatever these molecules and ions are doing on the atomic level.

Q Because the hydroxide ion is a strong base, it would not be totally unreasonable to think about an acid–base reaction between CH_3Br and OH^-. What would this reaction produce?

A As we will see later, there are two types of acid–base reactions. At this point you are probably most familiar with a Brønsted–Lowry acid–base reaction in which a hydrogen ion is transferred to the OH^- base. This transfer would result in the formation of water, which we should find as a product. But, in fact, we do not observe water as a product:

$$H^+ + OH^- \rightarrow H_2O$$ ■

We can also conclude that because the bromine of the methyl bromide is no longer attached to the carbon in the product, the reaction somehow involves breaking the carbon–bromine bond in the reactant.

Cleavage of a bond can occur in two ways: homolytically and heterolytically. *Homolytic* cleavage produces two neutral species because one of the electrons in the bond stays with one of the atoms, the other remains with the other. For example, the Br–Br bond usually breaks to give two Br• neutral atoms (*radicals*):

$$:\!\ddot{B}r\!-\!\ddot{B}r\!: \longrightarrow 2\;:\!\ddot{B}r\!\cdot$$

66 CHEMICAL REACTIVITY

Polar bonds, however, frequently cleave heterolytically by allowing the more electronegative atom to leave with both electrons:

$$H_3C\text{-}Cl \rightarrow H_3C^+ + Cl^-$$

Q What type of cleavage would be expected to occur in the C–Br bond?

A Heterolytic cleavage would be expected, leaving the carbon with a positive charge and the bromine with a negative charge. ∎

Thus, we begin to speculate, without any experimental evidence such as rates of reaction, that perhaps methyl bromide dissociates to give two ions—the CH_3^+ ion and the Br^- ion. It is natural for us to predict a heterolytic cleavage of the C–Br bond because one of the products is the stable bromide ion and because the carbon–bromine bond is polar. We further speculate that in order to form CH_3OH, a hydroxide ion must collide with the newly formed CH_3^+ ion, analogous to the manner in which a chloride ion collides with a silver ion to form solid AgCl. This bit of speculation is sensible because of the opposite charges of the two ions involved; the attraction of the ions should be considerable because of the favorable electrostatic interaction.

We have now developed a possible mechanism for the reaction. In a first step, a CH_3Br molecule dissociates to CH_3^+ and Br^-, and then in a second step, a hydroxide ion combines with the CH_3^+ ion to form CH_3OH:

$$CH_3Br \rightarrow CH_3^+ + Br^- \quad \text{first step}$$
$$CH_3^+ + OH^- \rightarrow CH_3OH \quad \text{second step}$$

Because the CH_3^+ ion goes on to react with hydroxide ion in a second step it is referred to as an *intermediate* in the reaction. The mechanism seems reasonable, but it is not yet sufficiently detailed nor has it been subjected to experimental sanction.

While we are still in the process of imagining or speculating, let's look at the first step again. Why would a CH_3Br molecule dissociate? Aren't these molecules stable? Surely, few of the CH_3Br molecules in fact dissociate in solution at room temperature. However, because this reaction is usually carried out in a polar solvent, there may be some energetic incentive for the dissociation to occur; that is, the ions will be attracted to the solvent molecules.

Q Why would a polar solvent such as water or an alcohol be attracted to the CH_3^+ ion?

A Polar molecules have an internal separation of charge that produces a dipole moment. The negative end of the dipole is attracted to the positively charged CH_3^+ ion. This attraction lowers the energy of the ion (in general, attractions lower the energy; repulsions increase the energy). The attractions to the CH_3^+ ion are shown below, but, of course, there is also the attraction of the positive end of the dipole (the tail of the arrow) to the bromide ion:

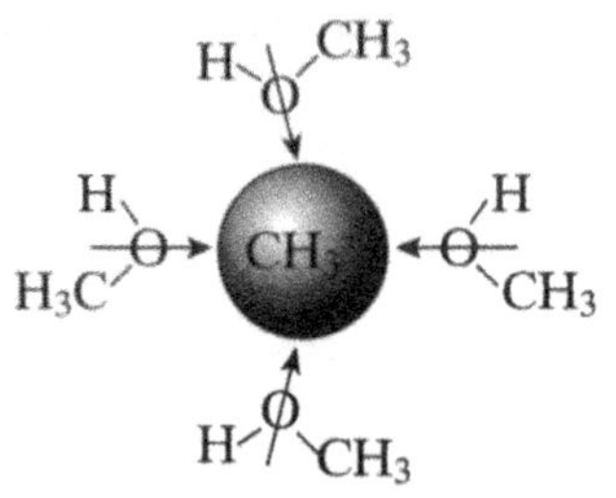

∎

Nevertheless, we must still imagine that energy would be required to cause the C–Br bond to break. Over the years, chemists have formulated models such as the collision model and transition state theory to describe these processes, and the term *activation energy* is used for the *minimum amount of energy that must be supplied to the reactants in order for the reaction step to occur*. In other words, we now look in more detail at breakage of the C–Br bond, and imagine the bond stretching and eventually reaching the point at which it will sever. The bromine will then carry away the electrons originally in the bond, and the carbon will be left with a positive charge. We suspect that the energy will be at a maximum when the C–Br bond is about to break; this energy is the activation energy for this step of the reaction.

It is useful to draw a graph of the energy of this step as a function of the distance between the carbon and the bromine. On the left side, the methyl bromide exists as a stable, unperturbed molecule; on the right side, the two ions have been generated and, assuming that they are in a polar solvent, are solvated.

The energy increases as the C–Br bond lengthens until the activation energy (E_a) is reached and the bond now has sufficient energy for the heterolytic cleavage to occur. During the cleavage, each of the atoms–C and Br–begins to develop charge, solvent molecules begin to surround each, and finally the ions exist "in isolation" as solvated species. The graph diagram that plots the energy change during this process is called a *reaction profile* and is shown in Figure 5.1.

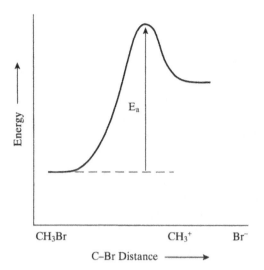

Figure 5.1. A reaction profile for the first step of a hypothetical mechanism.

So where does this energy, the activation energy, come from? We could, of course, simply heat the reaction mixture to try to provide enough thermal energy for the process, but in fact, chemists have long known that even at room temperature some molecules have fairly high energies. Indeed, there is a distribution of energies among molecules at any temperature. As the temperature is increased, a greater percentage of the molecules will have energies that are at or above the activation energy. This distribution is called the *Boltzmann energy distribution* and will be discussed in greater detail later in this chapter.

We now have speculated sufficiently about this particular mechanism and need to substantiate it with experimental work. The type of experimental work that can be used to support a mechanism will be the subject of the next several sections of this chapter. For now, we can tell you that the experimental work on the reaction of methyl bromide with hydroxide ion does *not* support this mechanism. In Chapter 6 we will examine some of this evidence and provide an alternative mechanism.

Q Many mechanisms are "explained" by showing how the electron flow might occur in a single step. Use the electron-pushing technique to show how the electrons in the C–Br bond wind up on the bromine in the first step of the mechanism above.

A

$$H_3C-\!\!\ddot{Br}\!: \longrightarrow CH_3^+ + :\!\ddot{Br}\!:^-$$

In this equation we show all of the lone pair electrons and have used an arrow to emphasize that the pair of electrons from the bond winds up on the bromine, thereby giving the bromide ion a total of four nonbonding pairs of electrons. ■

Q Now use electron pushing to show how the new C–O bond is formed.

A In this hypothetical mechanism the C–O bond is formed by reaction of the hydroxide ion with the CH_3^+ carbocation. The hydroxide ion has three pairs of nonbonding electrons, and one of these pairs is shared with the CH_3^+ ion as shown in the following scheme:

Note that the CH_3^+ ion is depicted with a trigonal planar structure (if you cannot predict this using the VSEPR model, please review that section in Chapter 3). ■

RATE OF REACTION

A chemical equation is a concise statement about a reaction that describes the reactants, the products into which they are converted, and the molar ratios pertaining to this conversion. It does not communicate anything about the rate at which the conversion occurs or the extent of the reaction. The reaction rate is often of great importance if the reaction is used as a preparative process or if the mechanism will be explored.

In general, the rate of a reaction is measured in terms of quantity per unit of time. Thus, for the reaction

$$A + B \rightarrow C + D$$

the rate might be expressed as the number of moles per liter of A (or B) consumed per minute or as the number of moles per liter of C (or D) produced per minute.

For this case

$$\text{Rate} = -\frac{\Delta[A]}{\Delta t} = -\frac{\Delta[B]}{\Delta t} = \frac{\Delta[C]}{\Delta t} = \frac{\Delta[D]}{\Delta t}$$

in which the symbol Δ represents a change for either a concentration or a time. The expressions for the reactants and products differ by a minus sign because the reactant concentrations decrease whereas the product concentrations increase.

For any given reaction the rate is greatly influenced by the specific conditions imposed on the reaction. Because chemical reactions involve the making and breaking of bonds, the rate depends on the specific bonds formed and broken, and therefore on the specific structure and composition of the reactants and products. Reactions that involve the combination of ions (e.g., aqueous acid–base neutralizations) generally occur very rapidly. Most nonionic reactions (e.g., organic chemical reactions) occur at considerably lower rates and require minutes or hours to reach equilibrium.

Concentration of the Reactants

For homogeneous chemical reactions, the reaction rate depends on the concentrations of one or more of the reactants. In general, the rate of a homogeneous reaction is proportional to the product of the molar concentrations of the reactants, each raised to some power. Thus, for the generalized reaction

$$aA + bB + \cdots \rightarrow$$

the reaction rate and the concentrations of reactants are related by the expression

$$\text{Rate} \propto [A]^m [B]^n$$

With the use of a proportionality constant k, this expression may be converted into the equation

$$\text{Rate} = k[A]^m [B]^n$$

This equation is the *rate law* for the specified reaction because it expresses the relationship between reaction rate and concentration of reactants under a specified set of conditions. The constant k is called the *specific rate constant* and is characteristic of that reaction with those specified conditions. Once the value of k has been determined for a particular reaction, the reaction rate may be predicted for any initial concentrations of A and B provided the specified conditions are maintained.

Statement. The exponents m and n in the generalized rate law are often referred to as the *order* of the reaction; m is the order with respect to substance A, n is the order with respect to B, and the sum $m + n$ is the overall order of the reaction.

Thus, a hypothetical reaction for which the rate law is

$$\text{Rate} = k[Q][R]^3$$

is first-order with respect to Q, third-order with respect to R, and fourth-order overall.

Statement. The values of m and n in the generalized rate equation cannot be predicted from the stoichiometry of the chemical equation; they must be determined experimentally.

Thus, a reaction equation describes the stoichiometry of the overall reaction, but it implies nothing about the order of the reaction with respect to any reactants.

As an example, consider the reaction of diatomic bromine with cyclohexene that yields *trans*-1,2-dibromocyclohexane:

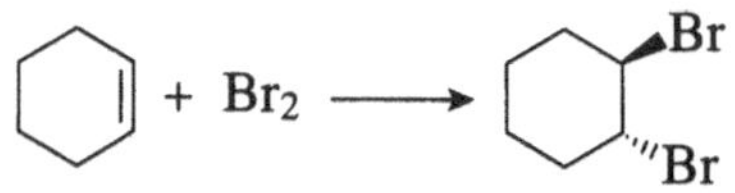

For this reaction at a specific temperature the experimental rate law is

$$\text{Rate} = k[Br_2]^2 [C_6H_{10}]$$

The powers to which the concentrations of bromine and cyclohexene are raised are not the same as their coefficients in the equation.

When the concentrations of bromine and cyclohexene are each 0.50 mol/L, the rate of formation of 1,2-dibromocyclohexane is 0.010 M/min (or mol/L·min). From these data the specific rate constant k at the given temperature can be calculated by using the expression for the rate law with the appropriate values; that is

$$0.010 \text{ M/min} = k(0.50 \text{ mol/L})^2 (0.50 \text{ mol/L})$$

and $k = 0.080 \text{ M}^{-2} \text{min}^{-1}$.

**Effect of Temperature on Rate:
The Arrhenius Equation**

Imagine a reaction that occurs by collision of two molecules. The number of collisions per minute that are effective in producing the product determines the rate of the reaction.

Q Is it true that whenever two molecules collide they form a new molecule?

A Collision is necessary to form a new molecule, but some collisions do not have sufficient energy to overcome the repulsions that exist between molecules at very close distances (see below). Also, the collision may occur with the molecules in orientations that do not align groups of atoms in a way that causes bonding. Consider the collision of the two A–B molecules shown below. If the products of the reaction are A–A and B–B, then it would appear that the atoms are not arranged correctly to allow the A group in one molecule to bond to the A group in the other molecule.

Therefore, few of the collisions that molecules undergo have sufficient energy or the proper orientation of the molecules to produce product.

Statement. The rate of a chemical reaction depends on the collision frequency, an energy factor, and a probability (orientation) factor.

The collision frequency depends on both concentration (or pressure for gases) and temperature. Increasing the concentration or the temperature will increase the rate of reaction. The probability factor depends on the shape of the reactants and the type of reaction. The most important factor is the energy factor, which depends on the energy of activation and the temperature of the reactant mixture.

The Boltzmann distribution of energy, shown in Figure 5.2, demonstrates how a temperature increase produces a greater proportion of reactant molecules that have a kinetic energy greater than (or equal to) the activation energy E_a.

Svante Arrhenius, a Swedish physicist (1859–1927), specified the mathematical relationship between the fraction of collisions with kinetic energy greater than the activation energy as $e^{-E_a/RT}$, in which R is the ideal-gas constant and T is the Kelvin temperature.

In the Arrhenius equation the exponential relationship indicates that a small difference in the activation energy has a large effect on the fraction of efficient collisions and hence the rate of reaction. In addition, raising the temperature will increase the reaction rate

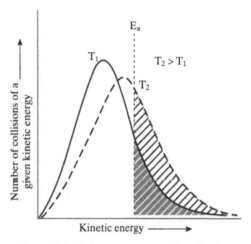

Figure 5.2. Boltzmann energy distribution.

by increasing the kinetic energy of the colliding molecules.

Statement. The Arrhenius relationship is generally represented in the following form for the rate constant k:

$$k = Ae^{-E_a/RT}$$

where A is the preexponential factor. The preexponential factor is also called the *frequency factor* and is related to the number of collisions and their orientation. In essence, then, the rate constant is obtained by multiplying the number of collisions by the fraction of effective collisions.

Q Explain why an increase in temperature increases the rate of any reaction independent of whether the reaction is endothermic or exothermic.

A An increase in temperature will increase the average kinetic energy of the molecules and increase the percentage of molecules whose energy is equal to or greater than E_a. Whether the reaction is endothermic or exothermic, overcoming E_a is the major factor in the *rate* of the reaction. The relative energies of the reactants and products, not the energy of activation, determine whether the reaction is exothermic or endothermic. We will discuss this in greater detail in the section on extent of reaction. ■

Determination of Rate Laws

One method for determining a rate law is to conduct a series of experiments for a specific reaction using

TABLE 5.3. Reaction Rate Data for Reaction of Cyclohexene with Hydrogen Chloride

Experiment Number	Initial Molar Concentrations		Molar Concentration of $C_6H_{11}Cl$ after 10 min	Rate (mol/L·min)
	C_6H_{10}	HCl		
1	0.100	0.100	0.015	1.5×10^{-3}
2	0.200	0.100	0.030	3.0×10^{-3}
3	0.300	0.100	0.045	4.5×10^{-3}
4	0.100	0.200	0.030	3.0×10^{-3}
5	0.100	0.300	0.045	4.5×10^{-3}
6	0.200	0.200	0.060	6.0×10^{-3}

different initial concentrations of one or more of the reactants. For example, consider the reaction between cyclohexene and hydrogen chloride under a given set of conditions that produces chlorocyclohexane:

$$\text{C}_6\text{H}_{10} + \text{HCl} \longrightarrow \text{C}_6\text{H}_{11}\text{Cl}$$

Suppose that a reaction mixture has equal concentrations (e.g., 0.100 M) of cyclohexene and hydrogen chloride, and the reaction proceeds for some convenient period of time, after which the mixture is analyzed to determine the concentration of chlorocyclohexane. After 10.0 min the concentration of chlorocyclohexane is 0.015 mol/L. The reaction rate, then, is 0.015 M/10 min = 1.5×10^{-3} mol/L·min. This experiment is repeated several times under identical conditions, except that in each run different initial concentrations of cyclohexene and/or hydrogen chloride are employed. Some results are given in Table 5.3.

The results of experiments 1, 2, and 3, in which the initial concentration of hydrogen chloride is constant, indicate that the rate is directly proportional to the molar concentration of cyclohexene. In other words, doubling [cyclohexene] doubles the rate and tripling [cyclohexene] triples the rate. Similarly, the results for experiments 1, 4, and 5, in which the initial concentration of cyclohexene remains constant, shows that the rate is also proportional to the molar concentration of hydrogen chloride. Therefore

$$\text{Rate} \propto [\text{cyclohexene}][\text{hydrogen chloride}]$$

and the rate law for this reaction is

$$\text{Rate} = k[\text{cyclohexene}][\text{hydrogen chloride}]$$

Q Are the results from experiment 6 in Table 5.3 consistent with this rate law?

A Yes, it is in agreement with the rate law because doubling both [cyclohexene] and [hydrogen chloride] increased the rate fourfold. ∎

Q One of your fellow students is confused. He thinks that experiment 6 shows that the rate should be proportional to [cyclohexene] + [hydrogen chloride]. Can you convince him that he is wrong?

A Assume that you start with 0.1 mol/L of cyclohexene and 0.1 mol/L of hydrogen chloride. If the rate were proportional to [cyclohexene] + [hydrogen chloride], then the rate would be proportional to 0.1 + 0.1 = 0.2 mol/L. Next start with 0.2 mol/L of cyclohexene and 0.2 mol/L of hydrogen chloride. The rate should now be proportional to 0.2 + 0.2 = 0.4 mol/L; that is, the rate would have doubled from the first experiment to the second. The data in Table 5.3 indicate that doubling the concentration of both reactants leads to a rate that is 4 times that of the first. Thus, the rate must be proportional to the product, not the sum, of the concentrations of the reactants. ∎

Q Consider the kinetic data in Table 5.4 for the formation of ethyl ethanoate (ethyl acetate) from ethanol and ethanoic acid (acetic acid) for a given set of reaction conditions. Determine the rate law.

A The rate is directly proportional to [ethanol] (experiments 1, 2 and 3), but doubling [ethanoic acid] increases the rate by a factor of 4 and tripling [ethanoic acid] increases the rate by a factor of 9 (experiments 1, 4, and 5); That is, the rate is proportional to [ethanoic

TABLE 5.4. Reaction Rate Data for $CH_3CH_2OH + CH_3CO_2H \rightarrow CH_3CO_2CH_2CH_3$

Experiment Number	Initial Concentration (mol/L)		Rate of Production of Ethyl Ethanoate (mol/L·min)
	Ethanol	Ethanoic Acid	
1	0.100	0.100	1.5×10^{-3}
2	0.200	0.100	3.0×10^{-3}
3	0.300	0.100	4.5×10^{-3}
4	0.100	0.200	6.0×10^{-3}
5	0.100	0.300	1.35×10^{-2}

acid]^2. The rate law, therefore, is

$$\text{Rate} = k[\text{ethanol}][\text{ethanoic acid}]^2 \quad \blacksquare$$

In both of the preceding examples, the numbers used for concentrations were chosen to make the relationship between concentration and rate as obvious as possible. In actual practice, however, any convenient concentrations can be used. If the rate law is not obvious from inspection of the tabulated data, it can be derived by calculations based on the generalized rate law

$$\text{Rate} = k[\text{A}]^m[\text{B}]^n$$

THE EXTENT OF REACTION: THERMODYNAMICS

Although both the rate constant and the equilibrium constant are measured experimentally, they are also related to the differences in certain intrinsic properties of the products and reactants of a reaction. These differences or changes are measured by variables that lie in the realm of thermodynamics, and include the *internal energy* E, the *enthalpy* H, and the *entropy* S, or *disorder*. Although there is a difference between E and H, for our purposes we will think of them as being related to the internal energy of the molecule. The enthalpy and entropy are related to another variable called the *free energy* G (or the *Gibbs energy*), $G = H - TS$, which in turn, is related to the equilibrium constant for a reaction $\Delta G° = -RT \ln K$, in which R is the ideal-gas constant and T is the absolute temperature. Note that the relationship between the free-energy change and the equilibrium constant involves $\Delta G°$, the change in the free energy measured under *standard conditions* [1 M concentration, 1 atmosphere (atm) pressure].

Q Rearrange the equation for $\Delta G°$ in terms of $\ln K$ and then determine whether a positive or negative $\Delta G°$ would produce an equlibrium constant greater than 1.

A The rearranged equation would be $\ln K = -\Delta G°/RT$. If the change in the standard Gibbs energy for a reaction is negative, the right side of the equation will be positive and $\ln K$ will be positive. If $\ln K$ is positive, K must be greater than 1 [remember that $\ln(1) = 0$]. Conversely, if $\Delta G°$ for a reaction is positive, K will be less than one. $\blacksquare$

If we can evaluate $\Delta G°$, we can obtain very valuable information about the extent of a reaction. As you might suspect, $\Delta G°$ is related to $\Delta H°$ and $\Delta S°$ when these thermodynamic values are measured under standard conditions:

$$\Delta G° = \Delta H° - T\Delta S°$$

Enthalpy changes are usually obtained from calorimetric measurements and are a measure of the *change* in electronic, vibrational, rotational, and translational energies as the reactants proceed to products. Most of the energy of a molecule is due to the electronic energy, which is a sum of the kinetic and potential energies of the electrons in a molecule and is primarily a result of the attraction of the electrons to the nuclei and the repulsions between all of the electrons. For energetically favorable reactions the potential energy of the products is lower than that of the reactants.

Statement. The change in ΔE (ΔH) is obtained by subtracting the energy of the reactants (E_r) from the energy of the products (E_p):

$$\Delta E = E_p - E_r$$

Therefore, an energetically favorable reaction ($E_p < E_r$) will have a negative ΔE and ΔH.

The *entropy change* is a measure of the change in disorder as the reaction proceeds:

$$\Delta S = S_p - S_r$$

Statement. If the products are more disordered than the reactants, then $\Delta S°$ will be positive.

Most of the disorder in a chemical reaction is produced by the translational motion of gaseous molecules. Therefore, a reaction has an increase in entropy when the number of molecules in the gaseous state is greater in the products than in the reactants.

Q Predict the sign of ΔS for the gaseous reaction

$$CO + 2H_2 \rightarrow CH_3OH$$

A For reactions that involve gases, the change in entropy is controlled by the difference in the number of gaseous molecules; that is, the 3 mol of molecules on the reactant side have a greater amount of disorder than the 1 mol of product. Therefore the ΔS for the reaction is negative because $S_r > S_p$. $\blacksquare$

Q Is $\Delta S°$ for the reaction of 1-butene with bromine in the gas phase negative or positive?

A In order to answer this question you need to write the equation for the reaction:

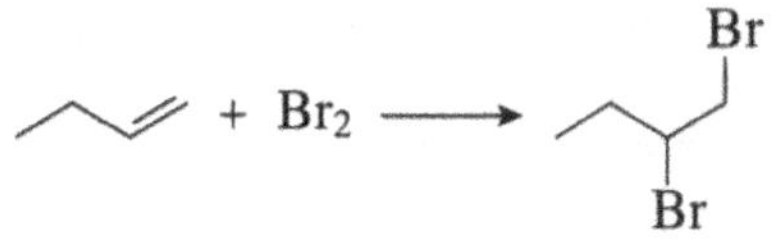

$$\text{CH}_2=\text{CHCH}_2\text{CH}_3 + \text{Br}_2 \longrightarrow \text{CH}_3\text{CHBrCH}_2\text{CH}_2\text{Br (with Br groups shown)}$$

Because the reaction converts 2 mol of reactant molecules into 1 mol of product molecules, the disorder should decrease and $\Delta S°$ should be less than zero. Indeed, for this reaction in the gas phase $\Delta S° = -142$ J/mol·K. ∎

In order to determine the extent of a reaction through $\Delta G°$, we need the values of $\Delta H°$ and $\Delta S°$. Fortunately, the following statement applies.

Statement. In many cases the magnitude of $\Delta H°$ is much larger than that of $\Delta S°$, and when that is the case, $\Delta G°$ is controlled by $\Delta H°$.

In the next section we will learn how to calculate a value for $\Delta H°$.

Calculation of $\Delta H°$

It is frequently important to be able to calculate the enthalpy change for a given reaction from known enthalpy changes in order to determine whether a reaction is favored energetically. This calculation is accomplished by utilizing the fact that enthalpy changes are additive. If two or more equations can be added, their enthalpy changes can also be added (and likewise for subtraction). If an equation is multiplied by some number, the enthalpy change, which is given in kilojoules per mole, is also multiplied by that same number. If an equation must be reversed, the enthalpy change for the reverse reaction is just -1 times the enthalpy change.

Q The enthalpy change for the dissociation of N_2 into atoms

$$N_2(g) \rightarrow 2\,N(g)$$

is 942 kJ/mol. How much energy is required to dissociate 56 g of N_2?

A The quantity of 56 g of N_2 is 2 mol of N_2, and therefore the energy required to dissociate this amount of N_2 is 2×942 kJ = 1884 kJ. Note that we are using the term *energy* as being interchangeable with *enthalpy*. ∎

Q How much heat is given off when 2 mol of nitrogen atoms in the gaseous state combine to form 1 mol of N_2?

A The combination of nitrogen atoms to form N_2 is just the reverse of the formula above for the dissociation of N_2:

$$2\,N(g) \rightarrow N_2(g) \quad \Delta H° = -942 \text{ kJ/mol}$$ ∎

Note that a negative enthalpy change corresponds to heat given off. This is also referred to as an *exothermic* reaction. A reaction in which the $\Delta H°$ is positive is an *endothermic* reaction; that is, heat is consumed by the reaction.

These relationships are known as Hess' law and are illustrated by the following example. Suppose that the standard enthalpy changes for the two reactions below are known

$$\text{CH}_4(g) + 2\,\text{O}_2(g) \rightarrow \text{CO}_2(g) + 2\,\text{H}_2\text{O}(g)$$
$$\Delta H° = -802 \text{ kJ/mol}$$

$$\text{CH}_2\text{CO}(g) + 2\,\text{O}_2(g) \rightarrow 2\,\text{CO}_2(g) + \text{H}_2\text{O}(g)$$
$$\Delta H° = -981 \text{ kJ/mol}$$

and we want to determine the enthalpy change for the reaction of methane with oxygen to form CH_2CO, the equation for which is

$$2\,\text{CH}_4(g) + 2\,\text{O}_2(g) \rightarrow \text{CH}_2\text{CO}(g) + 3\,\text{H}_2\text{O}(g)$$

Q Examine the two equations above and determine how to write them so that the CO_2 molecules will cancel *and* CH_2CO will be on the product (right) side of the formula.

A If we reverse the second equation so that CH_2CO appears on the product side of the formula, the two equations appear as

$$\text{CH}_4(g) + 2\,\text{O}_2(g) \rightarrow \text{CO}_2(g) + 2\,\text{H}_2\text{O}(g)$$
$$\Delta H° = -802 \text{ kJ/mol}$$

$$2\,CO_2(g) + H_2O(g) \to CH_2CO(g) + 2\,O_2(g)$$
$$\Delta H° = 981 \text{ kJ/mol}$$

Next, in order to cancel the CO_2 molecules, the first equation must be multiplied by 2, and then added to the second, which produces the desired equation:

$$2\,CH_4(g) + 4\,O_2(g) \to 2\,CO_2(g) + 4\,H_2O(g)$$
$$\Delta H° = 2(-802 \text{ kJ/mol})$$

$$2\,CO_2(g) + H_2O(g) \to CH_2CO(g) + 2\,O_2(g)$$
$$\Delta H° = 981 \text{ kJ/mol}$$

$$2\,CH_4(g) + 2\,O_2(g) \to CH_2CO(g) + 3\,H_2O(g)$$
$$\Delta H° = -623 \text{ kJ/mol}$$ ■

Enthalpy and Gibbs Energy of Formation

Enthalpy changes (and Gibbs energy changes) are frequently tabulated as standard enthalpies of formation. These are defined very precisely as the enthalpy change for the formation of the compound *from the elements in their most stable forms*, generally at 25 °C. For example, the standard enthalpy of formation of methane in the gaseous state is the enthalpy change for the reaction

$$C(s) + 2\,H_2(g) \to CH_4(g) \qquad \Delta H° = \Delta H_f°$$

Note that the gaseous methane is formed from graphite, the most stable form of carbon at 25 °C, and from elemental diatomic hydrogen gas, its stable form at 25 °C.

Statement. Because of the way in which enthalpies of formation are defined, the enthalpy change for a reaction is just the sum of the enthalpies of formation of the products, each multiplied by the coefficient in front of the product in the equation, minus the enthalpies of formation of the reactants, with their coefficients also taken into account.

For example, the enthalpy change for the reaction of carbon monoxide with hydrogen to form methanol

$$CO(g) + 2\,H_2(g) \to H_3COH(g)$$

is given by

$$\Delta H° = \Delta H_f°[H_3COH(g)] - \left\{\Delta H_f°[CO(g)] + 2H_f°[H_2(g)]\right\}$$

Statement. The enthalpy of formation of any element in its most stable form is, by definition, zero.

TABLE 5.5. Common Bond Energies (in kJ/mol)

H–H	432	C–F	485
H–C	411	C–Cl	327
H–N	386	C–Br	285
H–O	459	C–I	213
H–F	565	F–F	155
H–Cl	428	Cl–Cl	240
H–Br	362	Br–Br	190
H–I	295	I–I	148
C–C	346	C–N	305
C=C	602	C=N	616
C≡C	835	C≡N	887
C–O	358	N–N	167
C=O	709	N=N	418
C≡O	1072	N≡N	942

Use of Bond Energies

Approximate values for enthalpy changes can also be obtained from average bond energies. Bond dissociation energies can be measured for the dissociation of diatomic molecules such as H_2. The value of 432 kJ/mol is the amount of energy required for the process

$$H_2(g) \to 2\,H(g) \qquad \Delta H° = 432 \text{ kJ/mol}$$

The average or common bond energies listed in Table 5.5 are determined from dissociation energies in a large number of molecules and are consequently only an approximation for the bond energy in a given molecule.

The C–H bond energy of 411 kJ/mol can be applied to any reaction involving the homolytic dissociation of a C–H bond in the gas phase. For example, the energy required to break the four C–H bonds in gaseous methane, leaving one gaseous carbon and four gaseous hydrogen atoms, is 4 times the C–H bond energy:

$$CH_4(g) \to C(g) + 4\,H(g) \qquad \Delta H° = 4 \times 411 \text{ kJ/mol}$$

Even though these bond energies are not exact, they can be used to obtain an approximate value for the enthalpy change of many reactions in the gas phase.

Let's apply bond energies to the calculation of the enthalpy change for the hydrogenation of ethane:

$$H_2C=CH_2(g) + H_2(g) \to H_3C-CH_3(g)$$

In order to use bond energies we need to know which bonds have been broken in the reactants and which have been formed in the products.

 What bonds are present in H_3C-CH_3?

A There are three C–H bonds at each carbon and one C–C bond between the two carbons in ethane. ■

In the reactants, the C=C bond has been broken (or replaced with a C–C single bond) and the H–H bond has been broken. In the products, two C–H bonds have been formed and one C–C bond has been formed.

Q Add up the energies required to break bonds.

A The energy required to break a C=C bond and an H–H bond is 602 + 432 = 1034 kJ/mol. ■

Q Now, add up the energy released when the new bonds are formed.

A The energy released when the C–C bond and the two C–H bonds are formed is 346 + (2 × 411) = 1168 kJ/mol. ■

Q Finally, subtract the energy released by forming bonds from the energy required to break the bonds.

A 1034 − 1168 = −134 kJ/mol. ■

Thus, more energy is released when the bonds are formed in the products than the energy required to break the bonds in the reactants. Hence, the reaction is exothermic and releases 134 kJ every time a mole of ethane is hydrogenated when all reactants and products are in their standard states.

Q Use bond energies to estimate $\Delta H°$ for the reaction of methane with fluorine that yields difluoromethane. Is the following reaction exothermic or endothermic?

$$CH_4 + 2F_2 \rightarrow CH_2F_2 + 2HF$$

A $\Delta H° = \Sigma$ BDE for bonds broken − Σ BDE for bonds formed. [BDE = bond dissociation enorgy].

Bonds Broken	Energy (kJ/mol)	Bonds Formed	Energy (kJ/mol)
C–H	411	C–F	485
C–H	411	C–F	485
F–F	155	H–F	565
F–F	155	H–F	565
	1132		2100

$\Delta H° = 1132$ kJ − 2100 kJ = −968 kJ/mol

The reaction is exothermic because $\Delta H° < 0$. ■

TYPES OF REACTION

In your introductory chemistry course you learned about acid–base reactions, probably primarily about Brønsted–Lowry acid–base reactions. Acid–base reactions have been known since the time of the alchemists and are some of the most important types of chemical reactions. The various acid–base reactions differ in the way that the acid and base are defined. The two definitions of greatest importance in organic chemistry are the Brønsted–Lowry and the Lewis definitions.

Brønsted–Lowry or Proton Transfer Reactions

Statement. In the proton transfer or Brønsted–Lowry acid–base definition an *acid* is a proton donor and a *base* is a proton acceptor.

For example, acetic acid (CH_3CO_2H) will donate the proton (or hydrogen ion) of the OH group to ammonia, which serves as a base because its lone pair of electrons can accept the proton.

$$CH_3CO_2H + \ddot{N}H_3 \rightleftharpoons CH_3CO_2^- + NH_4^+$$

All Brønsted–Lowry acids must contain a hydrogen atom, and all bases must contain a pair of electrons that can be used to form a bond to the proton after it is transferred from the acid. The reaction between acetic acid and ammonia can be written as follows using an arrow to indicate the direction of *electron flow*:

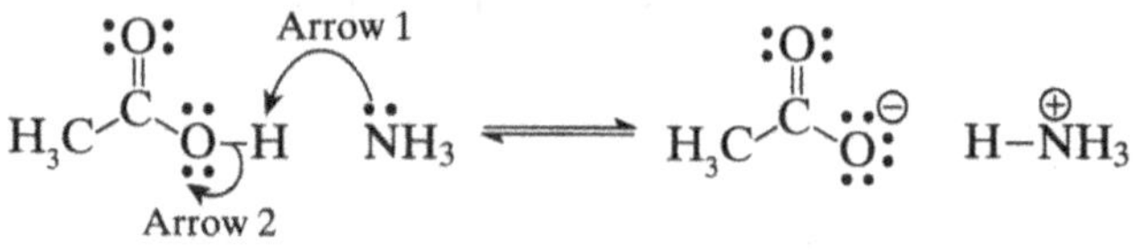

$$H_3C-C(=O)-\ddot{O}-H \quad \ddot{N}H_3 \rightleftharpoons H_3C-C(=O)-\ddot{O}:^{\ominus} \quad H-\overset{\oplus}{N}H_3$$

Arrow 1 as used above indicates that the lone pair of electrons on nitrogen will form a bond to the hydrogen ion as it is removed from the acid. Arrow 2 drawn from the O–H bond of the acid to the oxygen atom indicates the breaking of the O–H bond, leaving the pair of electrons in this bond on the oxygen. The Lewis structure of the acetate anion (product) shows that this pair of electrons gives the oxygen a formal negative charge.

Effect of Structure on Acidity and Basicity

Because we will frequently be concerned about the extent of the proton transfer process, we will need to review how structure affects the acidity of acids and basicity of compounds. First, however, it is important to remember that the extent of a reaction is generally described using an equilibrium constant. For acids it is denoted by K_a and for bases, the K_b. Organic chemists usually convert these constants to pK_a or pK_b.

These equilibrium constants are always defined for the reaction of the acid (or base) with water. Water is particularly convenient because it can function as both an acid and a base; that is, it is *amphiprotic* (think of the meaning of ambidextrous—the prefixes have the same origin). The amphiprotic nature of water results in its self-ionization:

$$H_2O + H_2O \rightleftharpoons H_3O^+ + HO^-$$

The equilibrium expression for this reaction is $K_w = [H_3O^+][OH^-] = 1 \times 10^{-14}$.

The K_a of 2×10^{-5} for acetic acid refers to the reaction

$$CH_3CO_2H + H_2O \rightleftharpoons CH_3CO_2^- + H_3O^+$$

and is associated with the equilibrium expression

$$K_a = \frac{[H_3O^+][CH_3CO_2^-]}{[CH_3CO_2H]}$$

If you look carefully at the equation showing the reaction of acetic acid with ammonia, you will recognize that every proton transfer reaction actually has two acids and two bases:

$$CH_3CO_2H + :NH_3 \rightleftharpoons CH_3CO_2^- + NH_4^+$$

Q What species transfer a proton (hydrogen ion)?

A On the reactant side of the formula, acetic acid transfers a hydrogen ion to the lone pair of electrons on the ammonia. On the product side of the reaction the ammonium ion transfers a hydrogen ion to the acetate ion in the reverse reaction that occurs as the reaction reaches equilibrium. If you turn the formula around, it is perhaps a little clearer that the ammonium ion is also an acid and the acetate ion is a base.

The acetate ion is referred to as the *conjugate base* of acetic acid, and the ammonium ion is the *conjugate acid* of the base ammonia. In other words, acetic acid and the acetate ion are a conjugate pair, and ammonia and ammonium ion are another conjugate pair. These conjugate relationships are important in establishing the relative extent of an acid-base reaction.

Q Write a formula for the reaction of HCl with triethylamine and then identify the conjugate pairs.

A

$$H-Cl + (CH_3CH_2)_3N: \longrightarrow (CH_3CH_2)_3N-H^+ + Cl^-$$

In the forward reaction a proton is transferred from H–Cl to the lone pair on the nitrogen of triethylamine. Thus, H–Cl is an acid and triethylamine is a base. In the reverse reaction, the triethylammonium ion is an acid (note that the lone pair is now tied up by being bonded to the hydrogen) and the chloride ion can function only as a base. Because the chloride ion is derived from HCl, the chloride ion is the conjugate base of the acid HCl. Because triethylammonium ion is derived from the base triethylamine, the triethylammonium ion is the conjugate acid of the base triethylamine.

$$H-Cl \text{ and } Cl^- = \text{conjugate pair}$$

$$(CH_3CH_2)_3N-H^+ + (CH_3CH_2)_3N: = \text{conjugate pair}$$

Let's look again at the reaction of ammonia with acetic acid, recognizing now that there are two acids and two bases in equilibrium. The position of the equilibrium—that is, whether the equilibrium favors the reactants or the products—depends on the extent of the hydrogen ion transfer on each side of the equation. If acetic acid reacts more "strongly" with ammonia than the ammonium ion reacts with the acetate ion, then more hydrogen ion transfers will occur between the reactants, thereby favoring formation of the products. In order to determine which acid, acetic acid or the ammonium ion, is the stronger one, we need to know the K_a for each species. We already know K_a for acetic acid (we can find listings of K_a values in any compilation of acid dissociation constants). We do not know K_a for the ammonium ion (and most compilations do not list the ammonium ion because it is the conjugate acid of the common base NH_3). The K_a value for an acid is related to the K_b value for its conjugate base through K_w, which is the self-ionization constant of water:

$$K_a = \frac{K_w}{K_b}$$

Q Determine K_a for the ammonium ion.

A K_w has a value of 1.0×10^{-14} at 25 °C and because K_b for ammonia is 2×10^{-5}, K_a for the ammonium ion is

$$K_a = \frac{1 \times 10^{-14}}{2 \times 10^{-5}} = 5 \times 10^{-10}$$

The K_a value of 5×10^{-10} ($pK_a = 9.3$) for ammonium ion is considerably smaller than the K_a of 2×10^{-5} ($pK_a = 4.7$) for acetic acid, and therefore acetic acid is a better donor of hydrogen ions. The following reaction will therefore favor the formation of the products and the extent of the reaction will be high:

$$CH_3CO_2H + \ddot{N}H_3 \rightleftharpoons CH_3CO_2^- + NH_4^+ \quad \blacksquare$$

Statement. In general, a proton (hydrogen ion) transfer reaction proceeds to minimize the concentration of the stronger acid and to increase the concentration of the weaker acid.

Although we can now qualitatively predict the extent of an acid–base reaction, in many cases it is necessary to determine its equilibrium constant.

Q Write the equilibrium expression for the reaction above and then show that the equilibrium constant can be found by dividing K_a for the reactant acid by K_a of the product acid (the conjugate acid of the reactant base).

A

$$K = \frac{[CH_3CO_2^-][NH_4^+]}{[CH_3CO_2H][NH_3]}$$

We know that the K_a value for acetic acid is given by the expression

$$K_a = \frac{[H_3O^+][CH_3CO_2^-]}{[CH_3CO_2H]}$$

and that the K_a of the ammonium ion is

$$K_a = \frac{[NH_3][H_3O^+]}{[NH_4^+]}$$

If we divide the K_a of acetic acid by the K_a of the ammonium ion, we obtain

$$K = \frac{[H_3O^+][CH_3CO_2^-][NH_4^+]}{[CH_3CO_2H][NH_3][H_3O^+]} = \frac{[CH_3CO_2^-][NH_4^+]}{[CH_3CO_2H][NH_3]}$$

The concentrations of hydronium ion cancel and the resulting expression is the correct one for the reaction. The equilibrium constant (K) for the reaction is therefore given by the ratio

$$K = \frac{K_a \text{ (reactant)}}{K_a \text{ (product)}} \quad \blacksquare$$

We now have the tools to predict the extent of a proton transfer reaction when we know something about the relative acidities or basicities of the reactants and products. Organic chemists frequently deal with molecules for which K_a or K_b values are not readily available, and it is very helpful to be able to predict their relative values. For example, we know that acetic acid is a weak acid in water, but what about methane? Is methane a proton donor? Methane has C–H bonds and therefore certainly can release a proton. But, in fact, the K_a for methane is very small (10^{-50}) and it is normally considered to be neutral.

Why is methane not acidic, but acetic acid is at least moderately acidic in water? In part, the relative proton-donating abilities of molecules are related to (1) the bond energy (strength) between an atom and the hydrogen atom to be removed and (2) the partial positive charge on the hydrogen atom. The greater the bond energy, the *weaker* the acid will be. The greater the partial positive charge on the hydrogen atom, the *stronger* the acid will be. For example, the energy required to break the S–H bond in H_2S is lower than the O–H bond energy in water, and it is therefore not surprising to find that H_2S is a stronger acid than H_2O—it is easier for a base to remove a proton by breaking the S–H bond. When molecules with the hydrogen attached to the same atom are being compared, the partial positive charge on the hydrogen atom influences the acidity. For example, chloroacetic acid and acetic acid both release a hydrogen ion from the O–H bond.

$$CH_3CO_2H + H_2O \rightleftharpoons CH_3CO_2^- + H_3O^+$$

$$ClCH_2CO_2H + H_2O \rightleftharpoons ClCH_2CO_2^- + H_3O^+$$

The electron density at the O–H hydrogen is different in the two molecules because of the electron-withdrawing effect of the chlorine. Chlorine has a greater electronegativity than the hydrogen for which it was substituted and therefore removes some electron density from the carbon to which it is attached.

Q Draw the structure of chloroacetic acid and show the charge separation in the C–Cl bond using the Greek symbol delta δ for a partial charge.

A

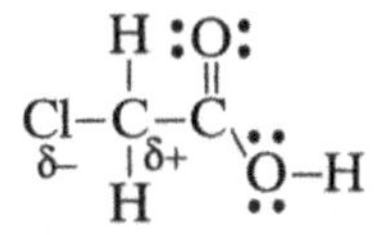

$$\underset{\delta-}{Cl}-\underset{\underset{H}{|}}{\overset{\overset{H}{|}}{C}}-\overset{\ddot{O}:}{\underset{\ddot{O}-H}{\overset{||}{C}}}$$

The alpha (α) carbon (the one next to the carbonyl) has a partial positive charge due to the withdrawal of electron density by the electronegative chlorine. This partial positive charge then affects the distribution of the electron density in the bond to the carbonyl group and induces a smaller partial positive charge on the carbonyl carbon. This small partial positive charge further induces a partial positive charge on the oxygen, which then slightly affects the distribution of electron density in the O–H bond. In the structure below, the smaller partial charge is indicated with two deltas, and an even smaller charge with three deltas:

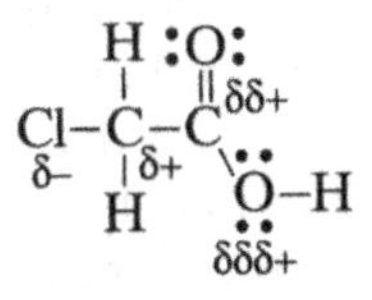

$$\underset{\delta-}{Cl}-\underset{\underset{H}{|}}{\overset{\overset{H}{|}}{C}}-\overset{\ddot{O}:}{\underset{\underset{\delta\delta\delta+}{\ddot{O}-H}}{\overset{||\delta\delta+}{C}}}$$

The overall effect of the chlorine is to produce a slightly greater partial positive charge on the hydrogen of the O–H bond. In picturesque terms this greater partial charge makes the hydrogen an easier target for the electron-rich site of the base. ∎

This polarization of bonds by a substituent is called the *inductive effect* and falls off rapidly the further the substituent is removed from the acidic hydrogen.

Q Which substituent would have the strongest electron-withdrawing inductive effect: F, Cl, Br, or I?

A Fluorine has the greatest electron-withdrawing effect because it has the highest electronegativity. ∎

Q Which acid in each of the following pairs is the stronger?

FCH_2COOH, $ClCH_2COOH$ FCH_2COOH, F_3CCOOH

A FCH_2COOH has a higher K_a (lower pK_a) than does $ClCH_2COOH$ because fluorine is a stronger electron-withdrawing substituent and therefore creates a larger partial positive charge on the H of the O–H bond. The F_3C group has three fluorines rather than the one fluorine on the FCH_2 group, and these remove more electron density from the hydrogen in the O–H bond. Therefore, F_3CCOOH is a stronger acid than FCH_2COOH. ∎

An important alternate perspective on the influence of substituents on relative acidities is to consider the stability of the conjugate base. Stabilization of the conjugate base means that it is more readily formed by the removal of a proton from the parent acid. Thus, the parent acid will be stronger if the negative charge in the conjugate base is stabilized by the inductive effect, by resonance, or by being dispersed over a larger volume. In chloroacetic acid the negative charge on the conjugate base is stabilized by the inductive effect of the chloro group. Therefore, stabilization of the conjugate base of chloroacetic acid also contributes to its greater acidity.

Q What about the *size* of the chlorine atom relative to a hydrogen? Is the conjugate base stabilized by the greater volume of the chlorine atom, which allows the electron density in the conjugate to be spread over a greater volume?

78 CHEMICAL REACTIVITY

A It is a quantum-mechanical fact that delocalization of charge reduces the energy of a system. Thus, the size of the chlorine also contributes to the stabilization of the conjugate base. ■

The electron density at acidic hydrogen atoms can also be affected by a substituent through resonance. This effect operates by delocalization of π electrons through part or all of the molecule. An example of the *resonance effect* is provided by the greater acidity of *p*-nitrophenol relative to the parent phenol.

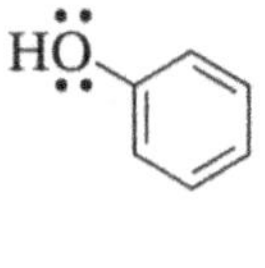

phenol
pK_a = 10.00

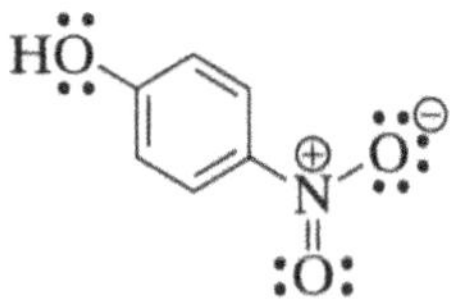

4-nitrophenol
pK_a = 7.15

A resonance structure that shows delocalization of electron density from the oxygen of the O–H group to the nitro group is shown below. This resonance form places a positive formal charge on the oxygen, giving the oxygen (and, in turn, the hydrogen) a greater partial positive charge.

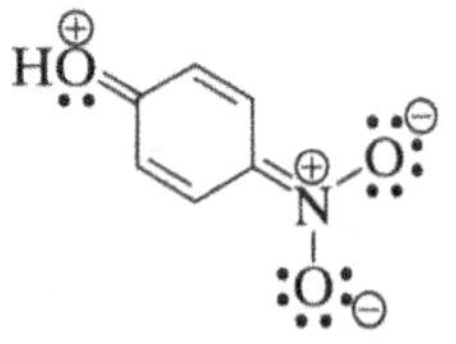

Q The resonance effect is not the only effect that increases the acidity of *p*-nitrophenol. The nitro group also exerts a significant inductive effect. Rationalize the electron-withdrawing inductive effect of the nitro group.

A In the Lewis structure

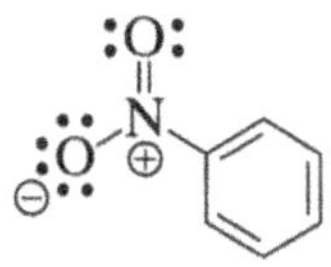

the nitrogen has a formal positive charge and is attached to two electronegative oxygen atoms. Because the NO$_2$ group is three bonds removed from the OH group in *p*-nitrophenol, the inductive effect is small and contributes little to an increase in the acidity of the compound. ■

Of course, we should also consider the relative stability of the conjugate base of *p*-nitrophenol rather than simply basing our analysis on the partial positive charge produced at the proton in the compound. The resonance structure below shows delocalization of electron density of the phenoxide ion to the nitro group, which lowers the energy of the conjugate base and increases the acidity of the nitrophenol.

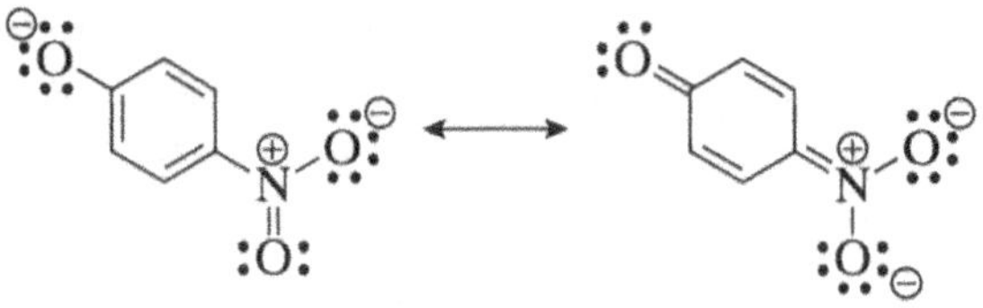

Resonance structures for conjugate base of 4-nitrophenol.

Q Start with the normal electron dot formula of *p*-cyanophenol and use electron pushing to derive a resonance structure similar to one for *p*-nitrophenol.

A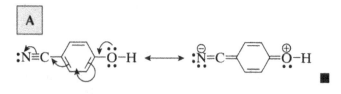
■

Q Nitromethane has a K_a of 6.3×10^{-11} (pK_a = 10.2). Account for the relatively high acidity of the methyl hydrogens by drawing the conjugate base of nitromethane. Is there any potential for charge stabilization?

A We can draw two resonance structures for the conjugate base of nitromethane. In the first the negative charge resides on carbon; however, in examining the second structure we can see that the conjugate base is stabilized by delocalization of the negative charge onto the more electronegative oxygen.

■

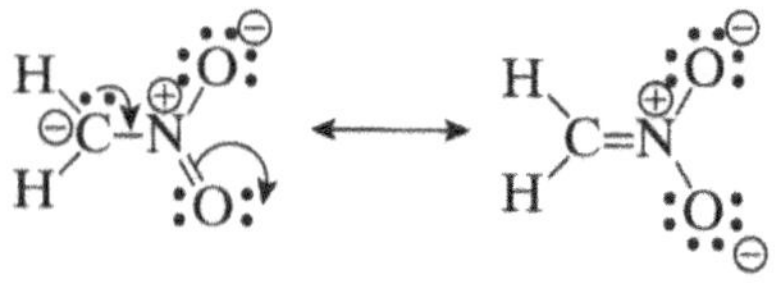

The examples above demonstrate the validity of the following statement.

Statement. All of the factors that affect the relative stabilities of the reactant *and* product must be considered in order to evaluate the extent of any reaction.

The *steric effect* is generally due to the presence of a large, bulky group close to the reactive site of a molecule. In many reactions the bulky substituent hinders the approach of a reactant, but for proton transfer reactions this effect is small because the hydrogen ion is so small. Bulky substituents do have an affect on acidity, however. For example, *o-t*-butylphenol

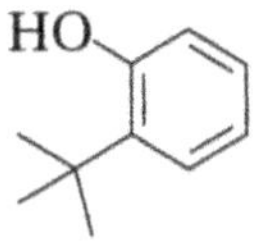

is considerably less acidic than phenol (the parent compound), but this effect is due mostly to the decreased interaction of the solvent (water) with the anion. Water interacts more strongly with the anion of phenol because of the attraction of the charge on the ion for the dipole of the water molecules (and hydrogen bonding), but with *o-t*-butylphenol the large *tert*-butyl group prevents close approach of the water molecules.

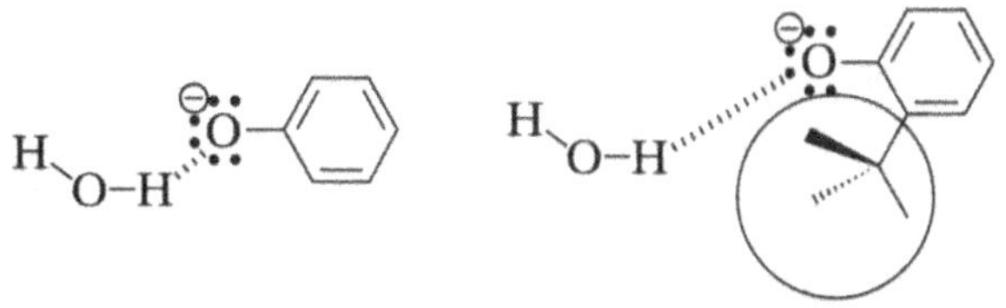

Q Can you think of a similar example in which the approach of the water molecule would be even more hindered?

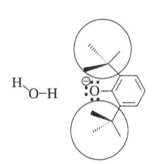

Proton Transfer Reactions in Organic Chemistry

In many organic reactions a hydrogen ion (a proton) will react with an electron-rich site on an organic molecule. These electron-rich sites usually contain a lone pair of electrons. For example, the oxygen of an alcohol contains two lone pairs of electrons that can be protonated.

$$CH_3CH_2\ddot{O}-H + H-A \rightleftharpoons CH_3CH_3\overset{+}{\underset{H}{\ddot{O}}}-H + :A^-$$

H–A represents a generic acid that is the source of hydrogen ions (protons).

In ketones and aldehydes, the carbonyl oxygen also has two lone pairs of electrons that can be protonated.

Q Use electron pushing to show the protonation of the lone pair of electrons on diethylamine by an acid HA.

A In amines, the nitrogen contains a single lone pair that is quite amenable to protonation (nitrogen is less electronegative than oxygen and therefore more basic).

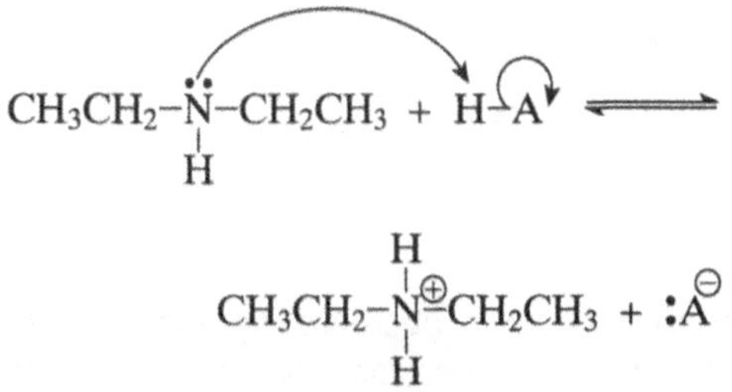

$$CH_3CH_2-\underset{H}{\overset{..}{N}}-CH_2CH_3 + H-A \rightleftharpoons$$

$$CH_3CH_2-\underset{H}{\overset{H}{\overset{|}{\overset{+}{N}}}}-CH_2CH_3 + :A^-$$

Q Write an equation showing the reaction of acetic acid with aniline (aminobenzene).

A

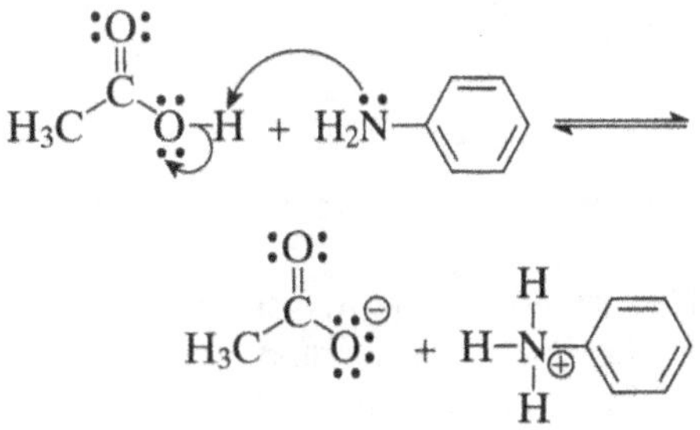

Note that every formula must be charge-balanced. In protonation reactions it is common to find that the basic site develops a formal positive charge, unless it is negatively charged. ∎

Q Compare the reaction above with the reaction of acetic acid with the phenoxide ion.

A

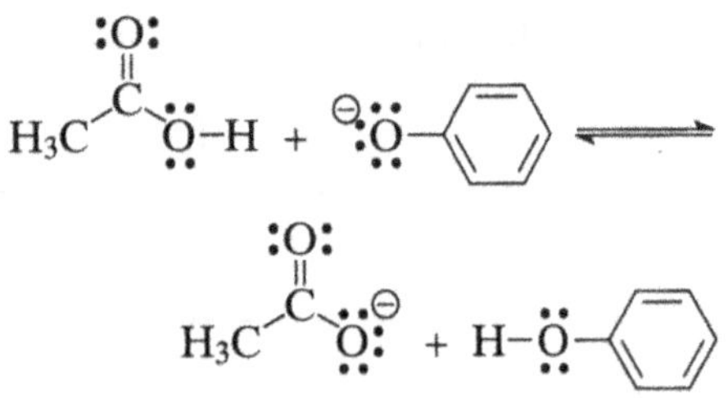

$$H_3C-C(=O)-O-H + {}^{-}O-C_6H_5 \rightleftharpoons H_3C-C(=O)-O^{-} + H-O-C_6H_5$$

∎

In the presence of base, some alkyl groups can lose a hydrogen ion. The site at which the hydrogen is lost is usually adjacent to an electronegative group. In the following example, a proton is abstracted from the carbon next to the electronegative carbonyl of ethanal.

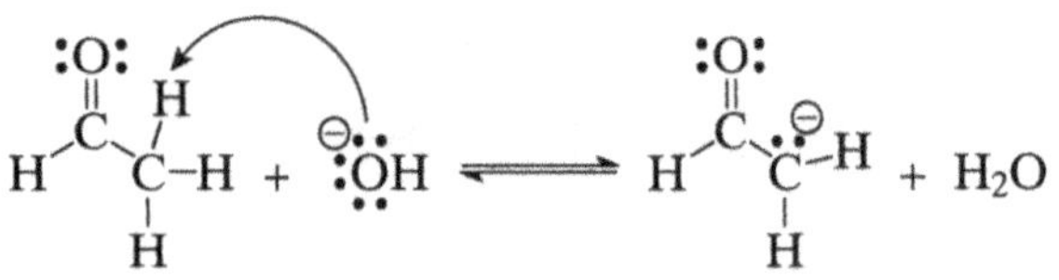

$$H-C(=O)-CH_2-H + {}^{-}OH \rightleftharpoons H-C(=O)-CH_2^{-} + H_2O$$

In some cases the hydrogen is made more acidic by the fact that the anion can be resonance-stabilized. The anion formed in the reaction of ethanal with base can exhibit the following resonance forms:

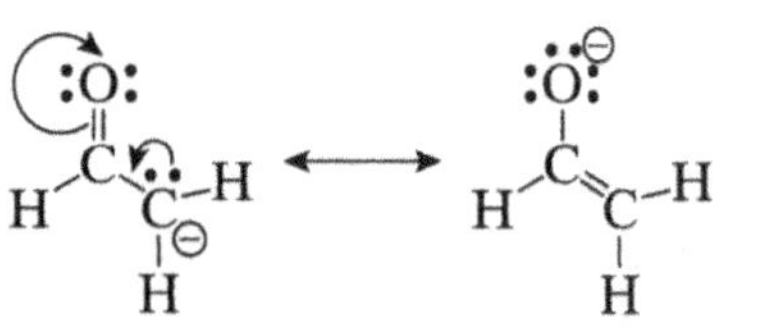

$$H-C(=O)-CH^{-}-H \leftrightarrow H-C(-O^{-})=CH-H$$

Q Determine which hydrogen will be removed from acetylacetone (2,4-pentanedione) and discuss the acidity of this compound relative to acetone.

A The compound will be more acidic than acetone because it has *two* electron-withdrawing carbonyl groups whereas acetone has only one.

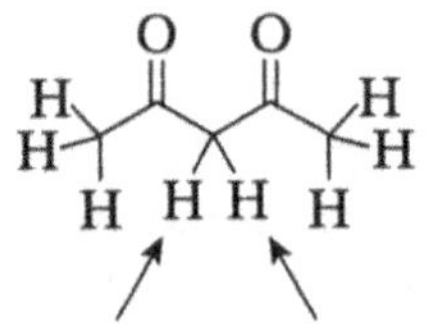

$$CH_3-C(=O)-CH_2-C(=O)-CH_3$$
(arrows point to the central CH$_2$ protons)

In this scheme, one of the protons marked with an arrow will be removed because they experience the electron-withdrawing effect of both carbonyl group and the negative charge on the conjugate base can be stablilized by resonance with both carbonyl groups. ∎

Q Use electron pushing to create resonance forms for the anion of acetylacetone.

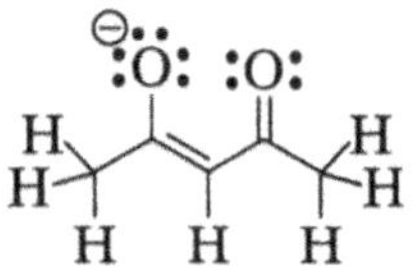

A

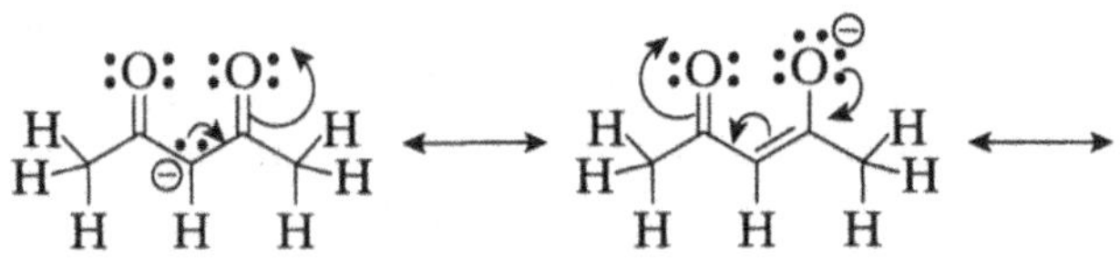

(a) $CH_3-C(=O)-CH^{-}-C(=O)-CH_3$ ↔ (b) $CH_3-C(=O)-CH=C(-O^{-})-CH_3$ ↔

(c) $CH_3-C(-O^{-})=CH-C(=O)-CH_3$

Note that if the movement of electrons in structure **a** goes to the left (rather than the right), the resonance form **c** is generated. ∎

Electron-Sharing or Lewis Acid–Base Reactions

The Lewis (the same G. N. Lewis of Lewis structure fame) definition of acids and bases was proposed at about the same time as the Brønsted–Lowry definition but is generally less familiar to students who have completed only general chemistry. The definition is more general (more encompassing) than the Brønsted–Lowry definition and also more useful, especially in organic chemistry.

Statement. According to the Lewis model, an acid is an electron pair acceptor. A base is an electron pair donor.

A typical Lewis acid–base reaction occurs between boron trifluoride and ammonia.

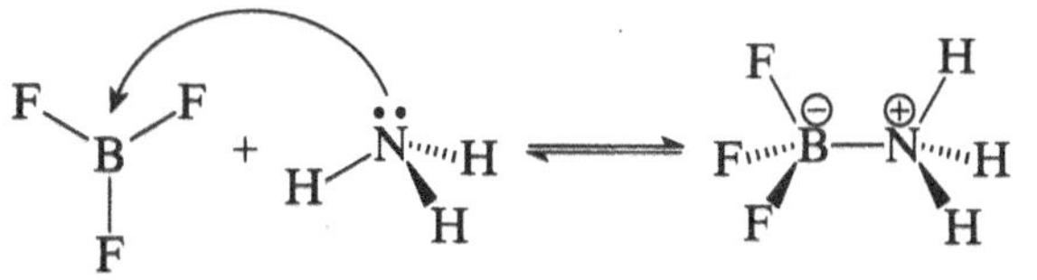

$$\underset{\text{Acid}}{BF_3} + \underset{\text{Base}}{:NH_3} \rightleftharpoons \underset{\text{Adduct}}{F_3B^{-}-N^{+}H_3}$$

Boron trifluoride is a Lewis acid because it lacks an octet of electrons on boron, which can thus accept an electron pair. The electron deficiency of BF_3 is emphasized by the Lewis structure (shown above), in which boron has only six electrons in its valence shell. Ammonia is a Lewis base because it contains a lone pair of electrons. The product of the reaction is called an *adduct* or *complex*. In the adduct the boron contains a negative formal charge because both electrons in the B–N bond have formally come from the nitrogen. Of course, the adduct is neutral and the nitrogen has a positive formal charge that balances the negative formal charge on boron.

You have probably noticed that ammonia, which we have used in the previous section as a Brønsted–Lowry base, is considered here to act as a Lewis base. In fact, all Brønsted–Lowry bases are also Lewis bases; the lone pair of electrons is common to both definitions.

Q Write an equation for the Lewis acid-base reaction of the common solvent tetrahydrofuran (THF) with BF_3.

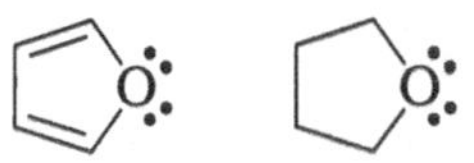

Furan   THF

A

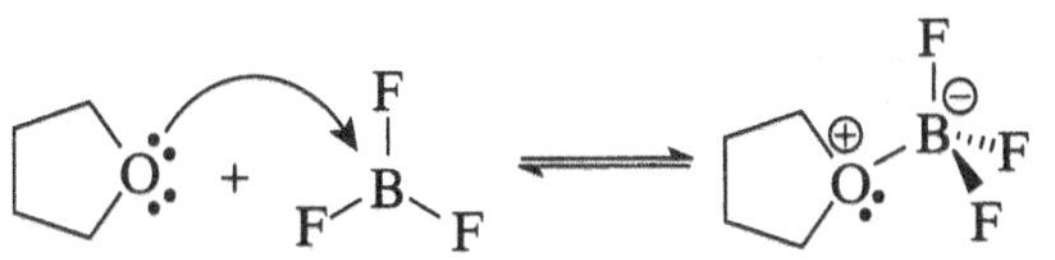

It is also true that all Brønsted–Lowry acids are also Lewis acids because proton donors donate their protons to electron pairs; that is, their protons are the electron pair acceptors. It is not true, however, that all Lewis acids are Brønsted–Lowry acids. Indeed, BF_3 contains no hydrogen atoms and therefore cannot be a Brønsted–Lowry acid.

Q Which of the following is more likely to be a Lewis acid—CCl_4 or Mg^{2+}?

A Carbon tetrachloride has the Lewis structure shown below, which indicates that the carbon has a complete octet of electrons.

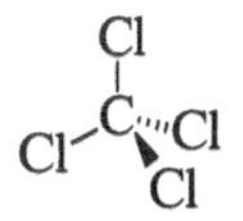

It could be argued that the electronegative chlorines surrounding the carbon remove enough electron density to give the carbon a partial positive charge and thereby make it a likely target for a Lewis base. However, the electrons donated by the base to the carbon must be able to enter an orbital; that is, the carbon must be able to accommodate them. The only orbital available on the carbon is an orbital in the quantum level ($n = 3$) above the quantum level of the valence shell ($n = 2$ for carbon). The $n = 3$ quantum level is at an energy that is too high to allow the carbon to use it.

The magnesium ion not only has a positive charge, which should attract electron density, but also has an empty orbital—the $2s$ orbital—that can be used to accept electron density. ∎

Indeed, a variety of chemical species can function as Lewis acids. Perhaps the most obvious species is a cation, which is positively charged and has room in its valence shell for a pair of electrons. The ammonia complexes made in the typical general chemistry laboratory are examples of Lewis adducts. For example, zinc ions react with four ammonia molecules to form the adduct or complex $Zn(NH_3)_4^{2+}$.

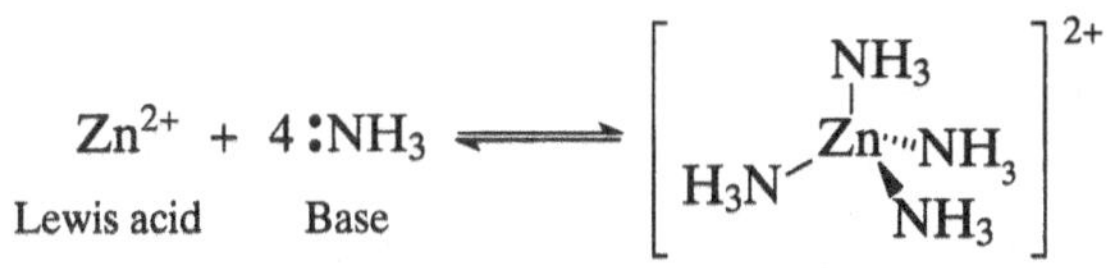

$$\underset{\text{Lewis acid}}{Zn^{2+}} + \underset{\text{Base}}{4:NH_3} \rightleftharpoons \underset{\text{Adduct}}{\left[Zn(NH_3)_4 \right]^{2+}}$$

Neutral molecules that can rearrange their electron density may also function as Lewis acids. For example, CO_2 reacts with hydroxide ion as shown below because the C=O bond can open to accept electron density from a base. Examine the diagram below to see how the electron flow occurs in this process:

$$\ddot{O}=C=\ddot{O} \quad {}^{\ominus}\!\!:\!\!\ddot{O}H \rightleftharpoons \overset{H\ddot{O}}{\underset{:\ddot{O}:}{{}^{\ominus}\!\!:\!\!C}}=\ddot{O}$$

Q Why would you expect CO_2 to react with the hydroxide ion even if you knew nothing about electron flow and electron pushing?

A The carbon of CO_2 has two electronegative oxygens attached to it and will therefore have a partial positive charge:

$$\overset{\delta-}{\ddot{O}}=\overset{\delta+}{C}=\overset{\delta-}{\ddot{O}}$$

In all of the cases mentioned so far, the atom that functions as the repository of the electron density has a partial positive charge. For example, the three electronegative fluorine atoms surrounding the boron in BF_3 give the boron a partial positive charge.

$$\delta-F\underset{F\delta-}{\overset{\delta+}{\diagdown B \diagup}}F\delta-$$

The same is true for the carbon of CO_2 and the tin of $SnCl_4$. Of course, the base, by virtue of its lone pair of electrons, has a partially negatively charged site. Consequently, analysis of Lewis acid–base reactions usually involves looking for sites with these characteristics. For example, the carbonyl carbon contains a partially positively charged carbon because it is attached to an electronegative oxygen atom.

Q Write a resonance structure for methanal that gives the carbon a formal positive charge.

A

$$\ddot{O}=CH_2 \longleftrightarrow {}^{\ominus}\!\!:\!\!\ddot{O}-\overset{\oplus}{C}H_2$$

Q Reason by analogy with the equation for the reaction of CO_2 with hydroxide ion to write an equation for the reaction of methanal with hydroxide ion.

A

$$\ddot{O}=CH_2 + {}^{\ominus}\!\!:\!\!\ddot{O}H \rightleftharpoons {}^{\ominus}\!\!:\!\!\ddot{O}-\overset{H}{\underset{H}{\overset{|}{C}}}-OH$$

Q Which of the following is not a Lewis acid–base reaction:

$$Zn^{2+} + 2OH^- \rightarrow Zn(OH)_2$$
$$Ag^+ + Cl^- \rightarrow AgCl$$
$$H^+ + H_2C=CH_2 \rightarrow H_3C-CH_2^+$$
$$Fe^{2+} + Zn \rightarrow Zn^{2+} + Fe$$

A In a Lewis acid–base reaction one reagent (the base) provides a pair of electrons to another reagent (the acid). In the adduct that results this electron pair is at least formally shared by two atoms. If the bonding is mainly ionic, as it is in $Zn(OH)_2$, the actual amount of sharing is relatively small. No matter what the actual extent of electron sharing, ion combination reactions such as the first two can be classified as Lewis acid–base reactions. The protonation of ethene is an example of a reaction in which the electron density provided by the ethene is definitely shared with the proton. The last reaction involves the *transfer* of electrons (not a *sharing* of electron density) from zinc metal to iron ions and thus is an oxidation–reduction reaction.

Q Write an equation for the reaction of $Al(OH)_3$ with hydroxide ion. Is it reasonable to think of $Al(OH)_3$ as a Lewis acid?

A
$$Al(OH)_3 + OH^- \rightarrow Al(OH)_4^-$$

$Al(OH)_3$ is amphoteric (reacts with both acid and base). When it reacts as a Brønsted–Lowry base, the hydroxide ion combines with a proton to form H_2O. Because the aluminum is electron-deficient (like the boron in BF_3), it can also function as a Lewis acid and react with hydroxide to form the $Al(OH)_4^-$ adduct.

Q The following reaction represents the protonation of ethene. Identify the Lewis acid, the base, and the adduct.

$$H^+ + H_2C{=}CH_2 \rightarrow H_3C{-}CH_2^+$$

A The hydrogen ion is the Lewis acid because it can and does accept electron density from the base, ethene. The product is the adduct, in which a pair of electrons, originally in the π bond of ethene, are now shared with the hydrogen ion. ∎

NUCLEOPHILES AND ELECTROPHILES

In organic chemistry the Lewis acid–base concept is generally applied to the *rates* of reaction. The term *nucleophile* (nucleus-loving) is used to indicate a species that is electron-rich and thus attracted to a site that is deficient in electron density. The electron-deficient site is called an *electrophile* (electron-loving). The terms *nucleophile* and *electrophile* parallel the terms *Lewis base* and *Lewis acid*, respectively, and are used in the context of reaction rates. Normally organic chemists are concerned with nucleophilic reactions that occur at a carbon atom.

Q Indicate all of the possible nucleophilic sites in the following molecules: 3-hydroxybutanal, methyl benzoate, *p*-methoxyphenol, and propyl amine.

A

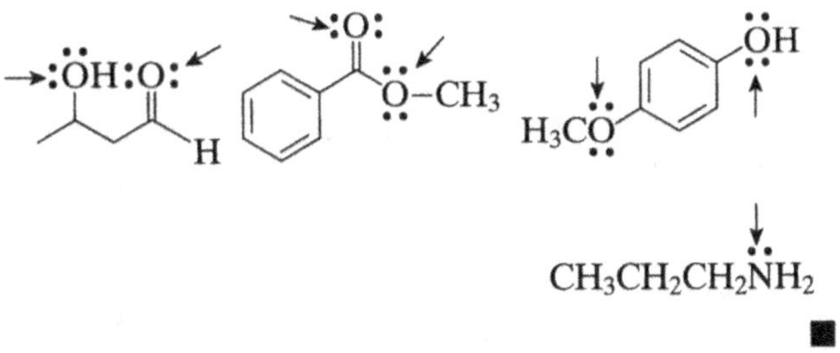

Q Indicate all of the possible electrophilic sites in the following molecules: 3-hydroxybutanal, oxirane, and chloroethane.

A

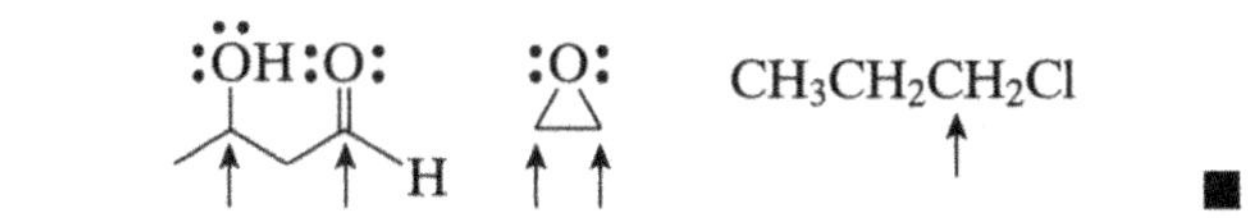

Let's look at two species, water and the hydroxide ion, that react as nucleophiles toward the methyl cation (the simplest example of an organic electrophile).

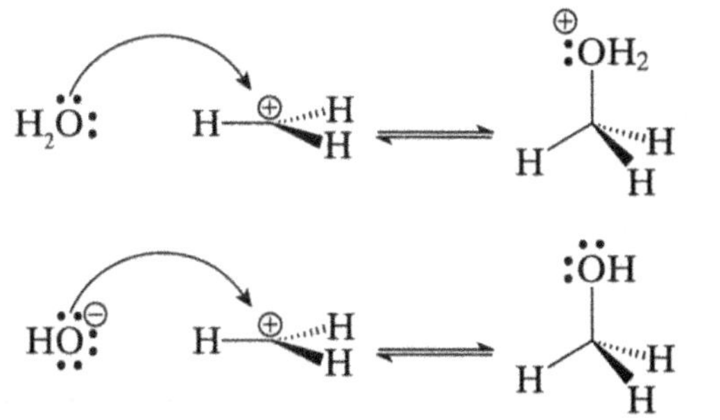

In the first reaction, water acts as a nucleophile and one of its lone pairs is donated to the carbon to form a bond to it. Note that the product is a protonated alcohol and that the total charge is conserved. (The requirement that charge on both sides of the equation must be the same frequently makes it easy to assign formal charges.) In the second reaction the hydroxide ion acts as a nucleophile and one of its lone pairs forms a bond to the carbon. Because hydroxide is negatively charged and the methyl cation is positively charged, the total charge on both sides of the reaction equation is zero. Of these two reactions the second one has the greater rate. The hydroxide ion is more nucleophilic than the water molecule because is has a negative charge that is more strongly attracted to the cationic center than is the water molecule, which is attracted only by an ion–dipole interaction. Hence, we say that hydroxide ion has a greater nucleophilicity (or is a stronger nucleophile) than water does.

A nucleophile can attack a neutral carbon atom if the carbon is somewhat electron deficient. Such is the case in the following reaction:

In this reaction, the carbon in methyl chloride is electrophilic (electron-poor) because the electronegative chlorine withdraws some electron density from the carbon. When a lone pair of electrons on the hydroxide forms a new bond to the carbon the carbon–chlorine

84 CHEMICAL REACTIVITY

bond must be broken in order to preserve the octet of electrons. The breaking of this bond is shown by the second arrow, which places the bonding pair of electrons on the chlorine.

Q Use arrows to indicate the flow of electrons in the reaction of methoxide ion, CH₃O⁻ with methyl acetate. Reason by analogy to reactions of other carbonyl compounds with a nucleophile.

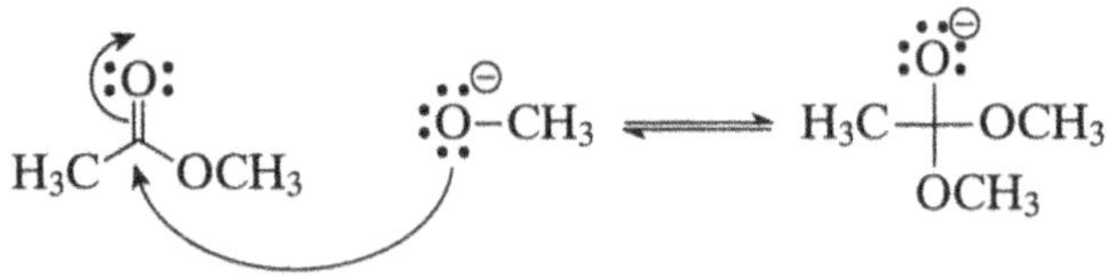

Q Use arrows to indicate the flow of electrons from ammonia to ethyl chloride. This is a substitution reaction in which ammonia substitutes for the chloride ion.

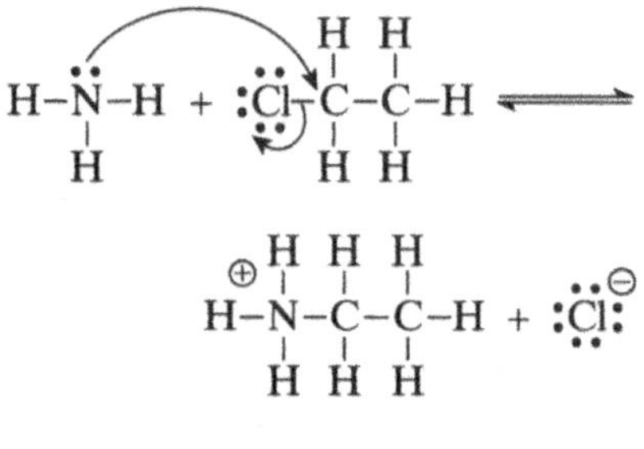

Q Use arrows to indicate the flow of electrons in the addition reaction of H₂S to the *tert*-butyl carbocation.

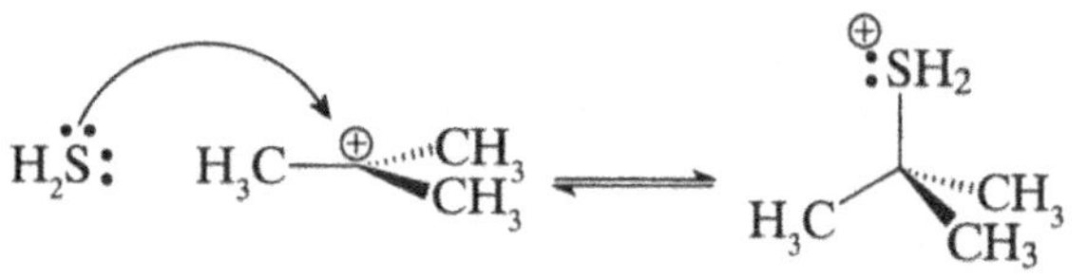

Statement. A basic principle in all the examples above is that electrons flow from the electron-rich site to the electron-deficient site.

Organic chemists are very careful to make a distinction between basicity and nucleophilicity.

Statement. The term *basicity* always refers to the extent of a reaction (i.e., the equilibrium position of a reaction), whereas *nucleophilicity* refers to the rate of a reaction.

For example, if we compare the reaction of methyl chloride with first hydroxide ion and then hydrogen sulfide ion (in ethanol as solvent), we find that the extent of the reaction with hydroxide ion is greater than the extent of the reaction with hydrogen sulfide ion; that is, hydroxide ion is the stronger base.

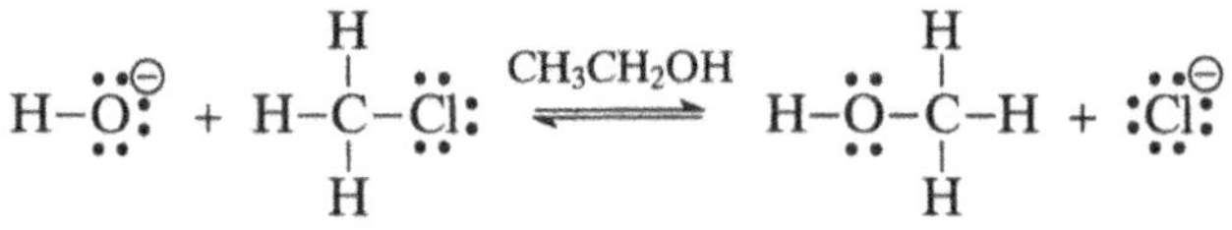

(greater extent of reaction)

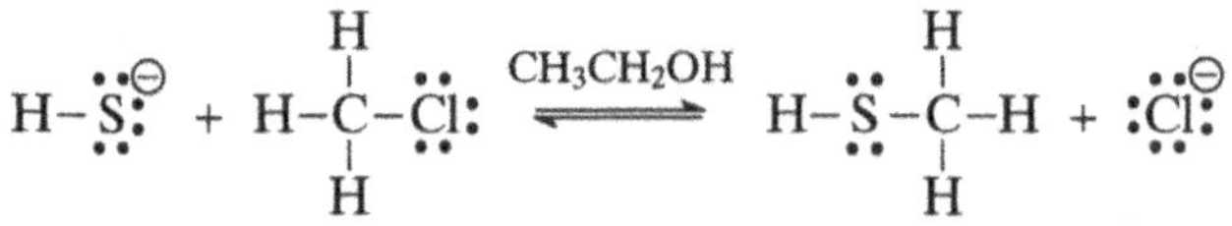

(greater rate of reaction)

However, the rate of the reaction with hydrogen sulfide is greater; that is, the hydrogen sulfide anion is a better nucleophile. Therefore, in this substitution reaction hydroxide has the greater basicity and lower nucleophilicity. We will discuss the effect of solvent on this nucleophilic substitution reaction in Chapter 6.

Q Consider the following two reactions and determine (1) which reactants are the nucleophiles, (2) why the nucleophiles attack the carbon of the carbonyl group, and (3) which reaction is faster.

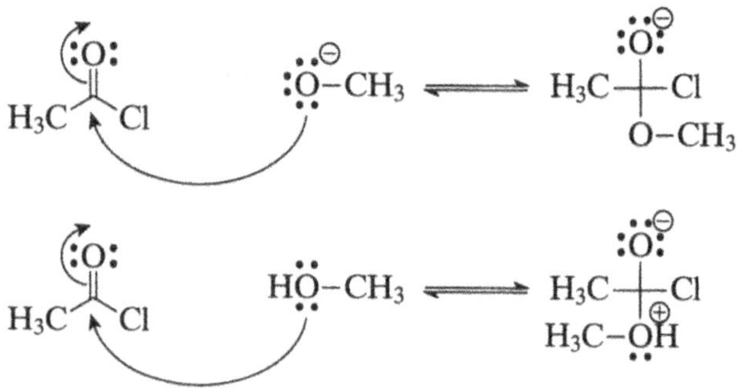

A (1) Methoxide ion is the nucleophile in the first reaction, and methanol is the nucleophile in the second reaction. (2) Nucleophilic attack occurs at this carbon because it is electron-deficient—the carbonyl carbon is attached to two electronegative atoms (oxygen and chlorine)

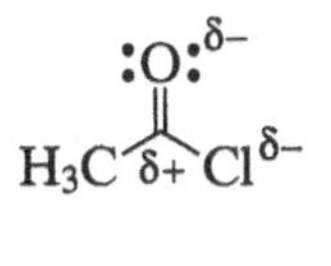

$$\text{H}_3\text{C}\overset{\overset{\displaystyle :\!\ddot{\text{O}}\!:^{\delta-}}{\|}}{\underset{\delta+}{\text{C}}}\text{Cl}^{\delta-}$$

and it will act as an electrophilic site. During the attack of the hydroxide ion on this carbon, the electron pair of the π bond of the carbonyl moves to the oxygen and establishes a formal negative charge there. Breaking the π bond is necessary to preserve the octet of electrons at the carbon. (3) The first reaction is the faster one because the negatively charged methoxide ion will be a better nucleophile than will the neutral methanol molecule. ■

6

REACTION MECHANISMS

A mechanism is a detailed, step-by-step molecular-level description of a reaction. It can be stated in words, but a structural explanation is better. Organic chemists usually rationalize the flow of electron density in each step by using electron pushing. Although a mechanism can be disproved if it is inconsistent with what is known experimentally about a reaction, it can never be proved. At most, we can say that a mechanism takes into account all empirical information about a given reaction.

Before we examine the types of evidence and reasoning that are used to propose a mechanism, we will identify the four major types of reactions.

REACTION TYPES

Organic reactions can be classified as substitutions, additions, eliminations, or rearrangements. The reaction of ethane with elemental chlorine to produce ethyl chloride and hydrogen chloride is an example of a *substitution reaction*. Note that one chlorine atom has been substituted for a hydrogen in ethane.

$$CH_3CH_3 + Cl_2 \rightarrow CH_3CH_2Cl + HCl$$

An *addition* reaction requires that a π bond be present in the organic reactant, which must therefore be an alkene or alkyne, or must contain a carbonyl group or other groups such as an imine (C=N). Hydrogen chloride adds to ethene to yield ethyl chloride.

$$H_2C=CH_2 + HCl \rightarrow CH_3CH_2Cl$$

In an *elimination* reaction a π bond forms with the elimination of part of the organic reactant as a second molecule. For example, heating ethanol in the presence of an acid catalyst such as sulfuric acid produces ethene and water.

$$CH_3CH_2OH \xrightarrow[\text{heat}]{H_2SO_4} H_2C=CH_2 + H_2O$$

Rearrangement reactions can be readily identified because the reactant and product are isomeric; the only chemical change that has occurred is the reorganization of bonds and atoms within the organic reactant. Exposure of 1-butene to an acid catalyst causes it to isomerize to 2-butene.

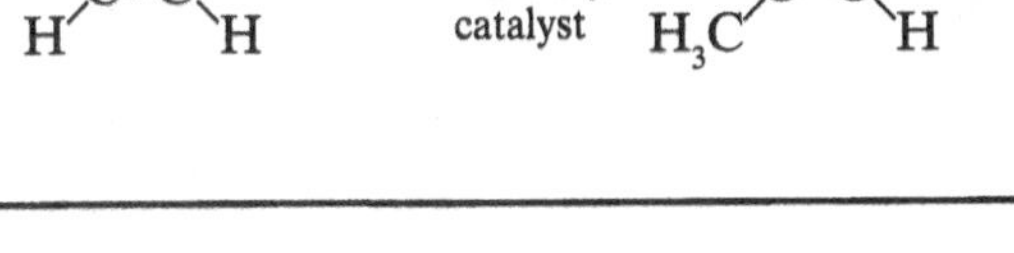

Q Identify the following reactions by their category:

$$CH_3O^- + CH_3CH_2Br \rightarrow CH_3CH_2OCH_3 + Br^- \quad (6.1)$$

$$CH_3CHO + H_2O \rightarrow CH_3CH_2(OH)_2 \quad (6.2)$$

A Reaction 6.1 is a substitution reaction with the methoxide ion substituting for bromine in a nucleophilic attack on ethyl bromide. Reaction 6.2 is an addition reaction with water adding to the carbonyl group.

The Bridge to Organic Chemistry: Concepts and Nomenclature
By Claude H. Yoder, Phyllis A. Leber, and Marcus W. Thomsen
Copyright © 2010 John Wiley & Sons, Inc.

Note that substitution and addition reactions generally have different forms. Substitution reactions by nature must expel a group and have the form A + B → C + D. Addition reactions do not expel a group and have the form A + B → C. ∎

BOND CLEAVAGE TYPES

Before we look at some examples of organic mechanisms, we also need to review the two types of bond cleavage. A bond can break either homolytically (in which one electron from the bonding pair goes with each fragment unit) or heterolytically (in which the bonding pair of electrons moves together). *Homolytic* cleavage tends to occur in bonds that have little or no bond moment. Diatomic chlorine has no dipole moment. Application of either sufficient heat or light of appropriate wavelength can cause the bond to break; when the Cl–Cl bond breaks, two chlorine atoms or radicals form. A *radical* is a species that has an unpaired electron. When a bond is broken evenly (homolytically), so that one elecron goes with one of the atoms and the other electron goes to the other atom, two "fishhook" arrows are used to show this process:

Cl–Cl ⟶ 2 Cl•   2 chlorine atoms or radicals

Q Explain why homolytic cleavage is much more common for the Cl–Cl bond than for the C–C bond.

A Remember that bond dissociation energy is the energy required to break a bond homolytically in the gas phase. Thus, the bond energy of 432 kJ/mol for H_2 corresponds to the energy required for the process

$$H_2(g) \rightarrow 2\,H\bullet(g)$$ ∎

The lower bond energy for the Cl–Cl bond can be rationalized in part by the difference in bond lengths. (In general, shorter bond lengths correlate with stronger bonds.) The carbon atom is considerably smaller than the chlorine atom, and consequently, the C–C single-bond length is shorter than the Cl–Cl bond length. However, bond length is not the only factor that affects bond energy. Bond order and ionic character are also important, and in Cl_2 the repulsion of the lone pairs on the adjacent atoms also seems to make the bond weaker than expected.

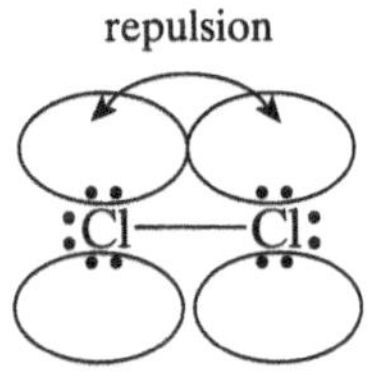

Repulsion of lone pairs of electrons.

Heterolytic cleavage, in contrast, occurs when a polar covalent bond breaks. In methyl chloride, for example, the polar C–Cl bond can break heterolytically. Because chlorine is the more electronegative atom in the bond, the bonding pair of electrons moves toward the chlorine to give the chloride ion. This process results in ions instead of radicals.

$$H_3C\text{–}Cl \longrightarrow {}^+CH_3 + Cl^-$$

MECHANISM OF HYDROGEN–CHLORINE REACTION

When a mixture of hydrogen and chlorine gases is exposed to light, a violent reaction occurs to form hydrogen chloride gas. The balanced equation for the reaction is

$$H_2 + Cl_2 \rightarrow 2\,HCl$$

In an attempt to propose a simple reaction mechanism, we might decide that a one-step, or concerted, mechanism seems reasonable. This mechanism would involve a cyclic transition state, as follows:

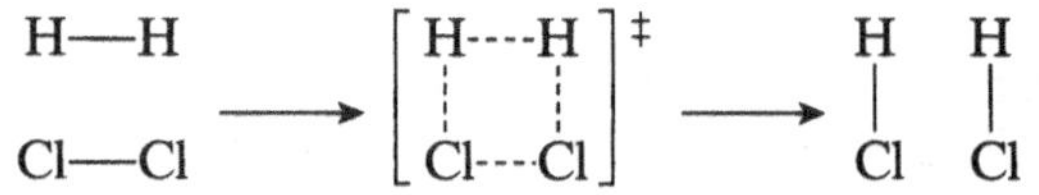

$$\begin{array}{c} H\text{—}H \\ Cl\text{—}Cl \end{array} \longrightarrow \left[\begin{array}{c} H\text{---}H \\ \vdots \quad \vdots \\ Cl\text{---}Cl \end{array} \right]^\ddagger \longrightarrow \begin{array}{cc} H & H \\ | & | \\ Cl & Cl \end{array}$$

Because both reactants are involved in a single step in a 1:1 ratio, the rate law for this mechanism would be

$$\text{Rate} = k[H_2][Cl_2]$$

If we examine the *actual* (experimental) rate law for the reaction of H_2 with Cl_2, however, we can exclude this mechanism. The reported rate law is

$$\text{Rate} = k_{\text{obs}}[H_2][Cl_2]^{0.5}$$

Having dismissed the concerted mechanism, we might ask whether diatomic hydrogen and chlorine are more prone to homolytic or heterolytic cleavage. Because both molecules are nonpolar, homolytic bond cleavage is favored. As we have seen above, the bond energy of Cl_2 is lower than that of H_2 and therefore the homolytic cleavage of diatomic chlorine is more likely. (In the following reactions the lowercase k over the arrows represents the rate constant for the given step.)

$$Cl\frown Cl \xrightarrow{k_1} 2\,Cl\cdot$$

Two chlorine atoms or radicals.

If one of the chlorine atoms (radicals) reacts with a molecule of diatomic hydrogen, a molecule of HCl can form:

$$H_2 + Cl\cdot \xrightarrow{k_2} H\cdot + HCl$$

The hydrogen atom (radical) can then react with another molecule of diatomic chlorine, which is much more abundant in the reaction mixture than atomic chlorine, to yield a second molecule of the HCl product.

$$Cl_2 + H\cdot \xrightarrow{k_3} HCl + Cl\cdot$$

We will not subject you to the details of the derivation of the rate law for this reaction, but, assuming that the dissociation of diatomic chlorine is reversible and rapid and that the second reaction is slower and therefore controls the rate of the reaction, we can express the rate law as

$$\text{Rate} = k_{obs}[H_2][Cl_2]^{0.5}$$

We are now ready to examine a related substitution reaction involving the organic hydrocarbon methane and diatomic chlorine.

CHLORINATION OF METHANE: A RADICAL MECHANISM

Chlorination of methane is a reaction that does not readily occur unless the reaction mixture of the two gases is heated in a sealed container or the container is exposed to ultraviolet light. The equation for the reaction is

$$CH_4 + Cl_2 \rightarrow CH_3Cl + HCl$$

The mechanism of the reaction that we propose must therefore result in these two products.

The wavelength of light required for the reaction corresponds to that known to cause bond dissociation of chlorine molecules. Because the bond dissociation energy of diatomic chlorine (240 kJ/mol) is less than that of a C–H bond (438 kJ/mol) in methane, the chlorination of methane is initiated by the heat- or light-induced homolytic cleavage of diatomic chlorine, as in the previous reaction.

$$Cl_2 \xrightarrow{\text{heat or light}} 2\,Cl\cdot$$

Q With what could a chlorine radical react?

A It could react either with a molecule of Cl_2 or CH_4 or another chlorine radical. The reaction with Cl_2 would simply produce another chlorine radical

$$Cl_2 + Cl\cdot \rightarrow Cl\cdot + Cl_2$$

and the reaction with another chlorine radical would just regenerate Cl_2, the reactant. However, when a chlorine atom collides with a methane molecule, a hydrogen atom from methane will be transferred to the chlorine atom to form HCl, which is one of the products.

$$CH_4 + Cl\cdot \rightarrow \cdot CH_3 + HCl \qquad \blacksquare$$

This step is the first of two propagations steps. A *propagation step* is a process in which a radical combines with a neutral species to generate another radical and another neutral species.

Q How is the product CH_3Cl generated?

A The reactive methyl radical could collide with a reactive chlorine atom to give this product, but there are many more chlorine molecules (Cl_2) than chlorine atoms (radicals) in the reaction mixture, so a methyl radical would more likely encounter a diatomic chlorine molecule.

$$\cdot CH_3 + Cl_2 \rightarrow CH_3Cl + Cl\cdot \qquad \blacksquare$$

Q Add the two propagation steps above to show that the sum is equivalent to the overall balanced reaction.

A Adding the two reaction equations and then eliminating species that occur as both reactants and products yields the appropriate overall reaction equation.

$$CH_4 + Cl\cdot \longrightarrow \cdot CH_3 + HCl$$

$$\cdot CH_3 + Cl_2 \longrightarrow CH_3Cl + Cl\cdot$$

$$CH_4 + \cancel{Cl\cdot} + Cl_2 + \cancel{\cdot CH_3} \longrightarrow$$
$$\cancel{\cdot CH_3} + HCl + CH_3Cl + \cancel{Cl\cdot}$$

$$CH_4 + Cl_2 \longrightarrow HCl + CH_3Cl \qquad\blacksquare$$

If the reaction mixture is exposed to UV light for a short time, the reaction proceeds through many cycles but eventually stops and reactive radicals recombine.

Q Think about what radicals have been formed during the initiation and propagation steps and then determine what compounds these radicals could form.

A In this *termination* of the reaction two chlorine atoms could recombine to give diatomic chlorine.

$$Cl\cdot + Cl\cdot \rightarrow Cl_2$$

Another possibility is that a methyl radical and a chlorine atom combine to give a methyl chloride molecule.

$$Cl\cdot + \cdot CH_3 \rightarrow CH_3Cl$$

It is also possible that two methyl radicals would combine to give ethane. Indeed, the observation of some ethane in the product mixture lends credence to this mechanistic proposal.

$$H_3C\cdot + \cdot CH_3 \rightarrow CH_3CH_3 \qquad\blacksquare$$

The fact that the reaction continues for many cycles even after the light source is removed can be rationalized by calculating the enthalpy of the gas-phase reaction.

Q OK, here is an opportunity to exercise your thermodynamic prowess by calculating the enthalpy change for the reaction of methane with Cl_2 using bond disocciation energies (BDEs).

A

$$CH_3\text{–}H + Cl\text{–}Cl \rightarrow CH_3\text{–}Cl + H\text{–}Cl$$
BDE(kJ/mol)   438   240   327   428

The sum of the bond energies on the reactant side is 678 kJ/mol; the sum of the reacting bond energies on the product side is 755 kJ/mol. This comparison shows us that the sum of the two new bonds on the product side is greater (i.e., the products have stronger bonds), and the reaction is therefore exothermic. Thus, $\Delta H°$ for the reaction is –77 kJ/mol (678 – 755). The substantial heat released by the reaction ensures that the reaction will be self-propagating for some period of time. $\blacksquare$

An *alternative* mechanism for the chlorination of methane is one in which the first propagation step is the reaction of a chlorine atom with methane to give methyl chloride.

$$CH_4 + Cl\cdot \rightarrow CH_3Cl + H\cdot$$

and a hydrogen atom. In the second step of this alternative mechanism the hydrogen atom would react with a molecule of diatomic chlorine to produce hydrogen chloride.

$$Cl_2 + H\cdot \rightarrow HCl + Cl\cdot$$

Note that the sum of these two propagation steps also yields the overall balanced reaction. The only way to differentiate between the two mechanisms is to examine the relative energetic requirements of the first step of each mechanism.

Below is the first step of the *first* mechanism.

$$CH_3\text{–}H + Cl\cdot \rightarrow \cdot CH_3 + H\text{–}Cl$$
BDE   438   428

In this mechanistic step the $\Delta H°$ is +438 – 428 = +10 kJ/mol. This mechanism therefore has a rate-determining first step that is modestly endothermic. [We will return to the discussion of rate-determining (slow) step later when we examine another multistep mechanism. For now, we will simply note that there can be only one rate-limiting step in a mechanism, and the identification of the rate-determining step must be consistent with the experimental rate law for a given reaction.]

In contrast, in the alternate mechanism the first step is considerably more endothermic: 438 – 327 = +111 kJ/mol.

$$CH_3\text{–}H + Cl\cdot \rightarrow H_3C\text{–}Cl + H\cdot$$
BDE   438   327

The *activation energy* for the rate-determining steps of these two mechanisms can be estimated from the enthalpy changes for each step. For the first mechanism, the activation energy should be relatively small because of the low enthalpy change (10 kJ/mol). By comparison, the activation energy for the alternate mechanism must be at least 111 kJ/mol. The higher predicted activation energy for the second mechanism suggests that it would be slower than the rate of the first mechanism and therefore probably does not compete with it.

Q Why does the activation energy for the alternate mechanism have to be at least 111 kJ/mol? Draw a reaction profile for the first propagation step of the two mechanisms.

A

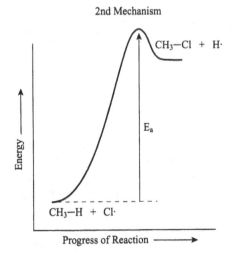

In Figure 6.1, note that E_a for the alternative mechanism is drawn considerably higher because the transition state is assumed to have an energy higher than that of the intermediate. It is also assumed that the transition states for both reactions are higher than the energies of the intermediates by about the same amount. This latter assumption is not necessarily correct. However, if this step determines the rate of the reaction, the activation energy must be greater than the difference between the energy of the intermediates and the energy of the reactants.

Next, we need to determine which propagation step in the first mechanism is the rate-determining step in the reaction.

$$CH_4 + Cl\cdot \rightarrow \cdot CH_3 + HCl \quad (6.3)$$

$$Cl_2 + \cdot CH_3 \rightarrow CH_3Cl + Cl\cdot \quad (6.4)$$

The fact that the first of the two propagation steps in this radical mechanism is the rate-determining step is consistent with the mechanism proposed for the reaction between hydrogen and chlorine gas. Further evidence comes from the fact that the experimental reactivity of diatomic halogens with methane decreases in the order $F_2 > Cl_2 > Br_2 > I_2$. This order definitely eliminates the dissociation of the diatomic halogen as a rate-limiting step because the bond dissociation energies for the halogens decrease in the order $Cl_2 > Br_2 > F_2 > I_2$.

Q Write all of the steps for the bromination of methane and calculate $\Delta H°$ for the reaction from $\Delta H°$ for each step.

A

$Br_2 \rightarrow 2\ Br\cdot$ initiation step

$Br\cdot + CH_4 \rightarrow \cdot CH_3 + HBr$ first propagation step
$\Delta H° = 76$ kJ/mol

$Br_2 + \cdot CH_3 \rightarrow CH_3Br + Br\cdot$ second propagation step
$\Delta H° = -135$ kJ/mol

$Br_2 + CH_4 \rightarrow CH_3Br + HBr$
$\Delta H° = -59$ kJ/mol

Figure 6.1. Reaction profiles for the first propagation step of the two methane chlorination mechanisms.

Note that ΔH for the initiation step is not taken into account because only a small amount of the radical is required to initiate the reaction.

Now we will examine another substitution reaction that occurs by a different mechanism—the reaction of methyl chloride with hydroxide ion to form methyl alcohol.

REACTION OF METHYL CHLORIDE WITH HYDROXIDE

Reaction as an Ionic (Polar) Mechanism

Let's first consider why the reaction of methyl chloride with hydroxide ion, a strong base, results in nucleophilic substitution rather than an acid–base reaction. The polar C–Cl bond in H_3C–Cl might produce sufficient electron deficiency at the hydrogens to make them somewhat acidic. The pK_a of chloroform ($CHCl_3$) is 29 ($pK_a = -\log K_a$) and the generally accepted value for the pK_a of methane is 50. If we assume that acidity effects are additive, then we can estimate the pK_a of dichloromethane (CH_2Cl_2) as 36 and that of methyl chloride (CH_3Cl) as 43.

Q Additivity relationships occur in many areas of chemistry. Let's go through the estimation of the pK_a of methyl chloride.

A If we assume that the effect of a C–Cl bond on acidity is independent of the nature of the molecule containing the bond, we can compare CH_4 with $CHCl_3$ and recognize that $CHCl_3$ has three C–Cl bonds, whereas CH_4 has no C–Cl bonds. Because the pK_a for methane is 50, whereas the pK_a for $CHCl_3$ is 29, we can assume that each C–Cl bond contributes $(50 - 29)/3 = 7\,pK_a$ units to the acidity. Thus, we estimate that the pK_a of CH_3Cl, with its one C–Cl bond, is $50 - 7 = 43$. ∎

The pK_a of 43 for chloromethane produces a very low equilibrium constant ($K_{eq} = 10^{-29}$) for the acid–base reaction of hydroxide and methyl chloride.

$$HO:^{\ominus} + H-\overset{H}{\underset{H}{C}}-Cl \rightleftharpoons H-\overset{\ominus}{O}-H + :\overset{H}{\underset{H}{C}}-Cl$$

Q Use the acid dissociation reactions for methyl chloride and water to verify that the equilibrium constant for the acid–base reaction is the ratio of the two acid dissociation constants.

A

$$CH_3Cl + H_2O \rightleftharpoons CH_2Cl^- + H_3O^+ \quad (K_a = 10^{-43})$$

$$OH^- + H_3O^+ \rightleftharpoons H_2O + H_2O \quad (K = 1/K_w = 10^{14})$$

Therefore

$$OH^- + CH_3Cl \rightleftharpoons H_2O + CH_2Cl^-$$
$$(K = K_a/K_w = 10^{-43} \times 10^{14} = 10^{-29})$$ ∎

Given this unfavorable acid–base reaction, we can assume that hydroxide ion does not remove a hydrogen ion from CH_3Cl. We can concentrate, therefore, on the nucleophilic attack of hydroxide on the carbon of methyl chloride. We have already argued that the electronegativity difference between carbon and chlorine produces a bond moment in which the chlorine atom is partially negative and the carbon atom is partially positive. Because the carbon is at the positive end of the bond moment, it is electrophilic. Hydroxide, which carries a negative charge, is a good nucleophile. Hence, the net reaction is the attack of hydroxide ion on methyl chloride to form methanol and chloride ion.

$$:\overset{\ominus}{\underset{..}{O}}H + H-\overset{H}{\underset{H}{C}}-\overset{..}{\underset{..}{Cl}}: \longrightarrow H\overset{..}{\underset{..}{O}}-CH_3 + :\overset{\ominus}{\underset{..}{Cl}}:$$

One mechanism for this reaction, considered briefly in Chapter 5, involves dissociation of the polar methyl chloride molecule to a methyl cation and a chloride ion.

$$H-\overset{H}{\underset{H}{C}}-\overset{..}{\underset{..}{Cl}}: \longrightarrow {}^{\oplus}CH_3 + :\overset{\ominus}{\underset{..}{Cl}}: \quad \text{Step 1}$$

$${}^{\oplus}CH_3 + :\overset{\ominus}{\underset{..}{O}}H \longrightarrow H\overset{..}{\underset{..}{O}}-CH_3 \quad \text{Step 2}$$

In this mechanism the first step involves heterolytic bond cleavage to form the methyl cation, and the second step of the mechanism is the reaction between the potent nucleophile hydroxide ion and the reactive electrophile methyl cation.

Statement. Generally, in mechanisms that consist of two or more steps, one of the steps is sufficiently slower than the other(s) that it controls the rate of the entire reaction.

Q If the proposed mechanism above is valid, which of the two steps is the probable rate-determining step?

A The slow step should be the first step because energy is required to break the carbon–chlorine bond. In contrast, the second step should be a fast step because of the high reactivity of the hydroxide anion and methyl cation, both of which have a full charge. In fact, most reactions between ions, such as the reaction of H^+ with OH^- or the reaction of Ag^+ with Cl^-, are fast because of the electrostatic attraction of the ions. ∎

If this were the actual mechanism and the first step, cleavage of the C–Cl bond, were rate-determining, the rate of the reaction should depend *only on the concentration of the methyl chloride*.

Q Table 6.1 presents some experimental data for reaction rate as a function of the concentrations of the reactants. Determine the rate law for the reaction.

A From Table 6.1 we can see that doubling the concentration of the methyl chloride doubles the rate of reaction. The same is true for the hydroxide ion reactant. Finally, doubling the concentration of both reactants leads to a fourfold rate increase. Thus, *experimentally* the rate of the reaction is proportional to the concentrations of both reactants; that is, the rate law is first-order in each reactant or second-order overall:

$$\text{Rate} = k[CH_3Cl][HO^-]$$ ∎

TABLE 6.1. Concentrations and Rates of the Reaction of CH_3Cl with OH^-

[CH_3Cl] (M)	[^-OH] (M)	Rate ($M^{-1}s^{-1}$)
0.05	0.10	0.6×10^{-5}
0.10	0.10	1.2×10^{-5}
0.05	0.20	1.2×10^{-5}
0.10	0.20	2.4×10^{-5}

Reaction as a One-Step Mechanism

The experimentally determined rate law for the reaction of CH_3Cl with OH^- indicates that both reactants are involved in the rate-determining step. Hence, we must revise our mechanism to account for the fact that the rate-determining step involves both the nucleophile and the electrophile. The simplest such mechanism would be a one-step (concerted) mechanism in which hydroxide would form a new bond to the methyl group in methyl chloride by displacing the chloride ion. We can designate this process as follows, using electron pushing to show that the electron-rich hydroxide will attack the electron-deficient carbon in methyl chloride.

$$HO\colon^{\ominus} + H-\underset{H}{\overset{H}{C}}-Cl\colon \rightleftharpoons HO-CH_3 + \colon Cl\colon^{\ominus}$$

Stereochemistry. In order to provide additional evidence for (or against) this mechanism, we next examine the *stereochemistry* of the reaction because the three-dimensional relationship of the orientation of nucleophiles and electrophiles is a very important aspect of most organic reactions. The stereochemistry of the reaction will depend on the way that the hydroxide ion attacks the methyl chloride. If the hydroxide attacks on the same side of the molecule that has the C–Cl bond, it is called *frontside attack*. If the hydroxide ion attacks on the opposite side of the C–Cl bond, it is termed *backside attack*.

Regardless of the type of attack, the chlorine must leave the molecule as the chloride ion. Thus, the chlorine is called the *leaving group*. The molecule that contains the electrophilic carbon is usually referred to as the *substrate*. Thus, ethyl chloride, propyl bromide, and isopropyl chloride are all substrates that could be subjected to this nucleophilic substitution reaction. The various groups to which chlorine is attached are frequently generically referred to by the symbol R. All the substrates containing chlorine could therefore be referred to as RCl.

Q Examine the reverse reaction and identify the nucleophile, the substrate, and the leaving group.

A In the reverse reaction

$$HO-CH_3 + \colon Cl\colon^{\ominus} \longrightarrow \:^{\ominus}\colon OH + H-\underset{H}{\overset{H}{C}}-Cl\colon$$

chloride ion would attack the electrophilic carbon in methanol. Thus, chloride is the nucleophile, methanol is the substrate, and the leaving group is the OH group. ∎

Now, we must determine the relative orientation of the nucleophile and the leaving group during the reaction.

Q Would the rate law for the reaction of hydroxide ion with methyl chloride depend on the orientation?

A No, because both nucleophile and substrate are involved in the only step of the reaction regardless of their orientations. ∎

Because this is a three-dimensional problem we use a substrate that will produce stereoisomers. Let's consider an alkyl chloride substrate that will permit an unambiguous determination of the stereochemical outcome of the reaction: 1-deuterio-1-chloroethane, which has a chiral carbon and therefore exists as a pair of enantiomers. Arbitrarily, we will start with the enantiomer that has the R configuration.

Q The R enantiomer is shown below along with the enantiomer formed by *frontside* attack. Assign the configuration of the product enantiomer.

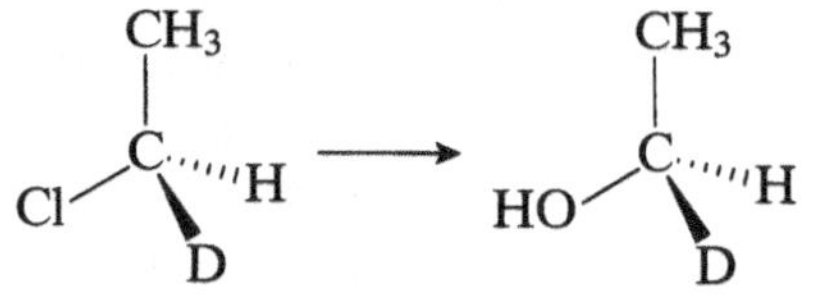

A The priority of the groups other than the hydrogen atom on this chiral carbon atom is O > C > D. An arc drawn to connect them in this order is a clockwise arc; thus the absolute configuration at this carbon is R. Both the reactant and the product have the R configuration. ∎

Frontside attack would therefore convert (R)-1-deuterio-1-chloroethane to (R)-1-deuterioethanol. Thus, if frontside attack occurs, the configuration of the product will be the same as that of the reactant.

The schemes below show that backside attack causes the same kind of inversion that occurs when an umbrella is blown inside–out. (Think of the C–Cl bond as the handle and the groups CH_3, D, and H as spokes in the umbrella. After inversion, the C–OH bond is the handle and the other groups have inverted their positions.) If backside attack occurs, the relative orientations of the groups will change—the R enantiomer of the reactant is converted to the S enantiomer of the product. Thus, the directional characteristics of the attack of the nucleophile result in a difference in the absolute configuration of the groups around the chiral carbon.

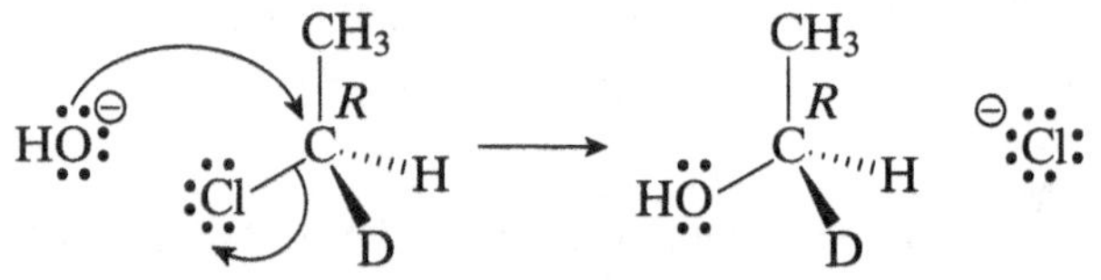

Frontside attack with retention of stereochemical configuration.

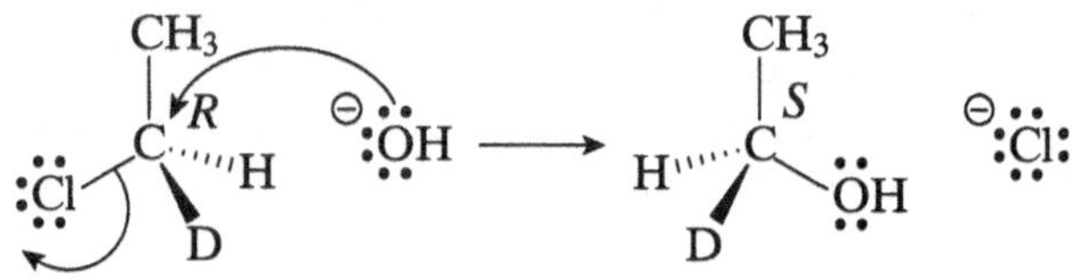

Backside attack with inversion of stereochemical configuration.

When this reaction is carried out in the laboratory, inversion of configuration is observed and therefore it appears that the backside attack mechanism is involved. Why would hydroxide ion attack the side opposite the carbon–chlorine bond? We might speculate that the negative charge on the nucleophile is repelled by the negative charge on the leaving group and that the electrostatic potential energy would be lowered by having the two as far apart as possible (see below).

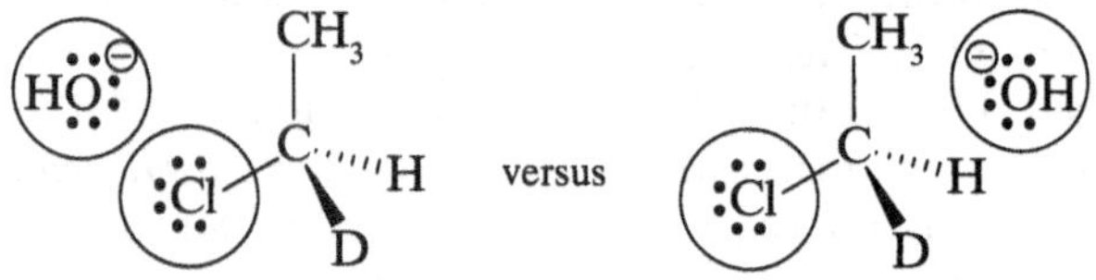

Backside attack minimizes electron–electron repulsions between the nucleophile and the leaving group.

A more sophisticated explanation involves the molecular orbital theory, the model of bonding that develops orbitals that encompass the entire molecule. These orbitals are the molecular analogs of atomic orbitals. A thorough discussion of the molecular orbital model is beyond the scope of this text, but the explanation that follows should suffice for the moment.

In our discussion of Lewis acids and bases, we found that the base (the nucleophile) contributes an electron pair (usually an unshared pair of electrons) to the Lewis acid, in this case the electrophilic carbon. Because this carbon already has eight electrons in its valence shell, there is no obvious orbital that can be used to hold the electrons from the nucleophile. Molecular orbital theory tells us, however, that each bonding orbital has an *antibonding* "partner." These antibonding orbitals are at a higher energy than the bonding orbitals but are not occupied by electrons and can therefore be used by the nucleophile.

The scheme below shows the bonding and antibonding molecular orbitals that are localized between the carbon and the leaving-group chlorine. The two types of shading represent the mathematical sign of the orbital. For example, the bonding orbital has a positive sign in the region between the two atoms where two electrons can provide the bonding that holds the two atoms together. In the antibonding partner, there can be no electron density between the two atoms (the small lobes have opposite mathematical signs). However, the antibonding molecular orbital has two areas that could accept electron density. One is located at the chlorine, where the nucleophile does not attack. The other is located at the carbon, where we know that the nucleophile does attack. The important thing to note is that this area is located away from the chlorine.

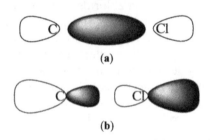

The σ-bonding (**a**) and σ*-antibonding (**b**) molecular orbitals at the C–Cl linkage.

Thus, the nucleophile directs electron density into the antibonding molecular orbital, away from the electron density of the C–Cl σ bond that is concentrated between the two atoms.

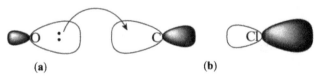

Overlap of a filled molecular orbital on the base with the empty antibonding C–Cl orbital: (**a**) nucleophile orbital; (**b**) σ*-antibonding orbital.

Effect of Leaving Group. We must now address the question of how the leaving group would affect the *one-step*, *concerted* mechanism. Because there is only one step in the mechanism, there can be only one transition state. The energy of this transition state relative to the energy of the reactants determines the activation energy, which in turn determines the rate of the reaction.

Q Draw a vertical line that represents relative energy and then place the energy of the reactants, the products, and the transition state on this line. Assume that the reaction is exothermic.

A

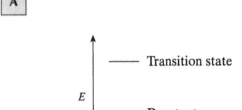

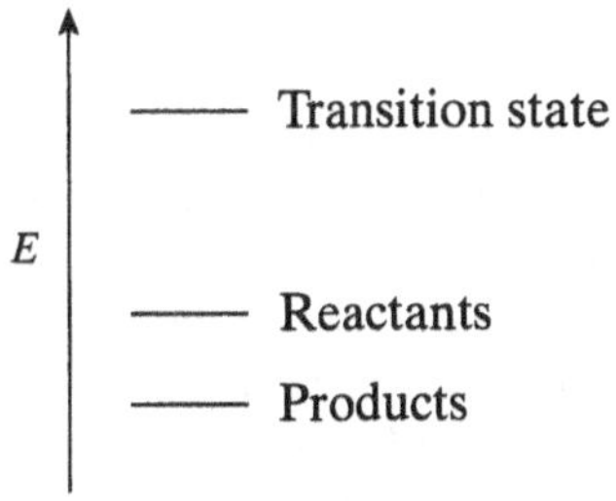

Because we assume that the reaction is exothermic, the energy of the products must be lower than that of the reactants (if this were not true, energy would not be released). The energy of the transition state must be higher than that of the reactants. ■

We now concentrate on the transition state. First, we must understand the following statement.

Statement. The transition state does not represent a compound that can be isolated; it is instead merely the highest energy that the reaction achieves in each step of the reaction. The transition state for each step in a mechanism does have structure, however, but this structure can only be inferred from factors such as the effect of the leaving group on the rate.

The transition state of the one-step, concerted mechanism can be imagined to have the structure shown below

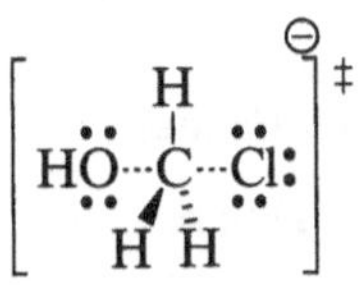

$$\left[\begin{array}{c} H \\ H\ddot{O}\cdots\overset{|}{\underset{H}{C}}\cdots\ddot{\ddot{C}l}\ddot{:} \\ H \end{array} \right]^{\ominus \ddagger}$$ possible transition state

where the oxygen of the nucleophile is beginning to form a bond to the electrophilic carbon, while the leaving group is beginning to break its bond to the carbon. The overall geometry of this assemblage of atoms is essentially trigonal bipyramidal.

Q Think carefully about this transition state and explain why it does not violate the octet rule.

A The three C–H bonds, which are normal two electron bonds, account for a total of six electrons. Because the bonds to the nucleophile and the leaving group are partial bonds, they can be assumed to involve only one electron each. The total number of electrons around the carbon is therefore $3 \times 2 + 1 + 1 = 8$. ∎

Finally, we can speculate about the effect of the leaving group on the energy of the transition state. Because the bond to the leaving group is being broken, we could certainly guess that the strength of the bond between the carbon and the leaving group would affect the energy. Because the leaving groups must eventually leave with a pair of electrons, the energy of the transition state is also lowered by the ability of the leaving group to delocalize and accommodate this charge. This ability is at least partly related to the size of the leaving group.

Q Given these two factors—bond strength and size—would you expect the C–I bond to produce the fastest rate for the methyl halides?

A Not only is the C–I the weakest of the carbon–halogen bonds (see Table 6.2), but the leaving-group iodide has the largest size of the halogens. Thus, iodide is the best leaving group (substrates with the C–I bond react more rapidly with a given nucleophile than do substrates with a C–F bond). ∎

Energetics, the Reaction Profile. Let's continue our analysis of the reaction of hydroxide ion with methyl chloride by considering the changes in energy that occur during the course of the reaction. The change in Gibbs free energy of the reaction is –91.2 kJ/mol ($\Delta G° = G_P - G_R$), and therefore the products are thermodynamically more stable than the reactants ($G_P < G_R$ because $\Delta G° < 0$). The $\Delta H°$ of the reaction, a function primarily of the bond strengths, is –75 kJ/mol. The exothermicity of the reaction results from the stronger C–O bond in the product relative to the C–Cl bond in the reactant. Although these thermodynamic data will help us construct a reaction profile for the reaction, they reveal nothing about how fast the reaction proceeds.

It is the height of the energy barrier (i.e., the activation energy) that controls the rate of the reaction. For the reaction of methyl chloride with hydroxide, a 0.05 M solution of CH_3Cl in 0.1 M NaOH(aq) at 25 °C undergoes only 10% reaction in 2 days. This result indicates that the activation energy is sufficiently high that relatively few molecules have sufficient energy at room temperature to overcome the energy barrier. The activation energy, determined using an Arrhenius plot of $\ln k$ versus $1/T$, is 105 kJ/mol. The thermodynamic and kinetic data can now be used to construct a *potential energy diagram* (*a reaction profile*).

Q A reaction profile shows how the energy varies as the reaction progresses. Create a reaction profile for the reaction of methyl chloride with hydroxide ion by plotting energy on the y axis, and on the x axis start with the reactants on the left side (at the intersection of the two axes) and place the products on the right side of the x axis. Imagine that you can measure the energy as a methyl chloride molecule collides (productively) with a hydroxide ion and then proceeds through the transition state to the product methanol and a chloride ion.

A Refer to Figure 6.2. ∎

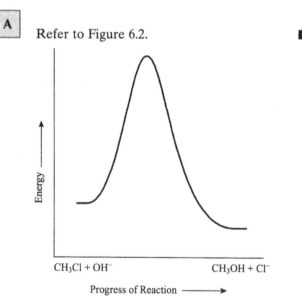

Figure 6.2. Reaction profile for the reaction of methyl chloride with hydroxide ion.

TABLE 6.2. Data Related to Methyl Halides

Methyl Halide	Electronegativity (EN) of Halogen	Difference in EN (C vs. X)	C–X Bond Strength (kJ/mol)
CH_3–F	4.0	1.5	460
CH_3–Cl	3.0	0.5	351
CH_3–Br	2.8	0.3	293
CH_3–I	2.5	0.0	234

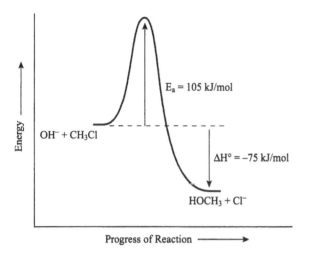

Figure 6.3. More quantitative reaction profile for reaction of CH_3Cl with OH^-.

In drawing a reaction profile it is generally a good idea to establish the relative energies of the products and reactants. Because the reaction is exothermic, the products are more stable than the reactants. The next step is to know whether the reaction has only one step or a series of steps. This reaction (methyl chloride reacting with hydroxide ion) is a one-step reaction, and therefore as the two reactant molecules approach one another the electrons of the nucleophile begin to repel the electrons around the electrophilic carbon until sufficient energy is gained to reach the transition state. In the transition state the electron–electron repulsions are overcome and a bond between the nucleophile and the carbon begins to form and, at the same time, the bond to the chlorine begins to break. The activation energy is the energy difference between the transition state and the reactants. As the C–O bond is more firmly established and the C–Cl bond length continues to increase, the system becomes more stable and the energy decreases until methanol and the chloride ion are completely formed. The reaction profile is shown again in Figure 6.3 in a more quantitative fashion.

Q Summarize the evidence for the concerted (one-step) mechanism.

A (1) The second-order rate law (first-order in nucleophile and first-order in electrophile) indicates that a reasonable mechanism is a one-step, concerted reaction that involves both reactants in the rate-determining step; (2) inversion of stereochemical configuration at the electrophilic carbon in the substrate suggests that backside attack of the nucleophile on the substrate occurs; and (3) a rate dependence on the bond strength of the carbon–halogen bond provides evidence for a transition state that involves bond making to the nucleophile and bond breaking of the carbon with the leaving group. ∎

Effect of the Solvent. Up to this point we have said nothing about the solvent for the reaction. Most organic reactions occur in a solvent because the solvent dissolves the reactants and allows for rapid diffusion of the molecules and relatively rapid rates of reaction. Polar solvents may be *protic* or *aprotic*. *Protic* solvents possess acidic hydrogen atoms that may form hydrogen bonds, while *aprotic* solvents lack acidic hydrogen atoms and thus cannot form hydrogen bonds.

Q Let's be clear about the meaning of "acidic hydrogen atoms." Which of the following have acidic hydrogen atoms? Which are protic solvents?

Hexane, methanol, acetic acid, ethyl acetate

A When we discuss solvents, the term *acidic hydrogen* refers to a hydrogen attached to an oxygen or nitrogen. Methanol has an O–H hydrogen that is acidic relative to the hydrogen atoms in hexane, but methanol is about 10 orders of magnitude less acidic than acetic acid. Ethyl acetate does not have a hydrogen attached to an oxygen or nitrogen, and therefore is not protic. Ethyl acetate, like methanol and acetic acid, has a dipole moment and therefore is polar. All polar solvents have dipole moments and interact with other dipolar molecules by the *dipole–dipole* interaction shown below. They also interact with ions by the even stronger *ion–dipole* interaction. ∎

Dipole–dipole interaction (negative end of dipole is attracted to positive end of neighboring dipole)

Ion–dipole interaction (negative end of dipole is attracted to positive ion)

Most protic solvents form strong hydrogen bonds to species with lone pairs of electrons. The diagram below shows hydrogen bonding between methanol and the hydroxide ion.

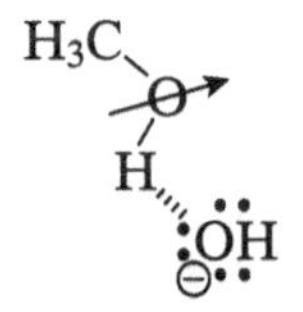

$$H_3C-\overset{..}{\underset{H}{O}}\cdots\overset{..}{\underset{..}{O}}H^{\ominus}$$

Hydrogen bonding between OH of methanol and hydroxide ion (H interacts with a lone pair of electrons on hydroxide ion)

For the reaction between hydroxide ion and methyl chloride, the reaction rate will be greater in a polar aprotic solvent than in a polar protic solvent. For example, if N,N-dimethylformamide (DMF) and methanol are used as solvents for this reaction, the rate of this reaction will be 100,000 times greater in DMF than in methanol. DMF cannot form a hydrogen bond to hydroxide ion but interacts with the hydroxide ion by ion–dipole interactions. Methanol interacts with the hydroxide ion by hydrogen bonding in addition to ion–dipole interactions. This relatively strong hydrogen-bonding interaction reduces the reactivity of the hydroxide ion by lowering its energy. In more picturesque terms, the H bonding forms a significant sphere of solvation that the hydroxide ion must escape before reacting with methyl chloride. In contrast, the nucleophilicity of the hydroxide ion is greater in DMF because it does not have to break H-bonding interactions with the solvent before attacking the methyl chloride.

Q One point that we have stressed is that the extent of a reaction is determined by the relative energies of the reactants and products, while the rate of a reaction is determined by the difference in energy of the reactants and the transition state in the rate-determining step of the reaction (the activation energy). The paragraph above seems to indicate that we need only examine the effect of the solvent on the reactant. Is this correct?

A We must always examine the energy of *both* the reactants and transition state in order to predict or rationalize relative rates. An increase in the solvent polarity actually decreases the reaction rate because the potential energy of the charge-dense hydroxide ion is reduced *more* by the polar solvent than is the potential energy of the transition state, which has the charge dispersed. Thus, the energy of activation for this reaction is greater in a polar solvent.

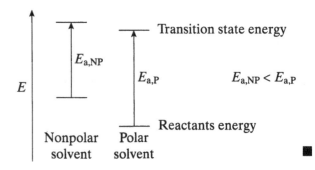

The S_N2 Mechanism. The concerted, one-step mechanism that we have now finished exploring is usually called an S_N2 or *bimolecular nucleophilic substitution mechanism*. The S_N indicates substitution, nucleophilic while the 2 tells us that two molecules are involved in the rate-determining step of the reaction. Of course, in this mechanism there is only one step and only one transition state. In the transition state the HO–C bond is partially formed and the C–Cl bond is partially broken so that overall the carbon still has approximately the same electron density that it would have in a stable tetravalent carbon compound. The mechanism for the S_N2 reaction of hydroxide with methyl chloride is shown below.

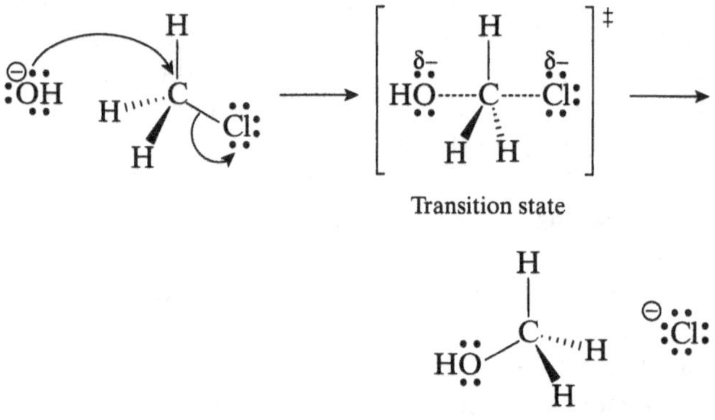

The net negative charge in the S_N2 transition state is dispersed over three atoms (O, C, and Cl), although greater charge density is carried by the electronegative oxygen and chlorine atoms. Dispersal of charge in the transition state relative to the charged hydroxide reactant helps us understand the effect of solvent polarity on the rate of this reaction.

Reaction as a Mechanism with a Trigonal Bipyramidal Intermediate

We conclude our discussion of this substitution reaction by briefly considering an *alternative* mechanism that is also consistent with the second-order rate law. In this mechanism the first bimolecular step consists of the reaction of hydroxide with methyl chloride to form a trigonal bipyramidal intermediate. This intermediate is improbable and highly energetic because it violates the Lewis octet rule by surrounding the carbon with 10 electrons and gives carbon a formal charge of −1.

Improbable trigonal bipyramidal intermediate.

The second step would then involve loss of the chloride to form the neutral methanol product.

Q Write equations for this mechanism.

A

$$CH_3Cl + OH^- \rightarrow [HO-CH_3-Cl^-] \quad (6.5)$$

$$[HO-CH_3-Cl^-] \rightarrow CH_3OH + Cl^- \quad (6.6)$$

Statement. Note that brackets are used to designate an *intermediate*, which is an unstable species that can be detected, and at times, even isolated. A transition state, on the other hand, may be designated by a bracket with a double dagger [‡]. A transition state, unlike an intermediate, is the highest-energy, hypothetical structure attained by the reactants in a particular mechanistic step. The transition state can be neither detected nor isolated.

While violation of the octet rule and the negative charge on carbon in the trigonal bipyramidal intermediate severely disadvantage this mechanism, let's at least consider whether it is consistent with all the empirical (experimental) evidence for the reaction of CH_3Cl with OH^- and other nucleophiles.

Q In order to see why the rate of this alternative mechanism would be unaffected by the nature of the halide, consider the effect of the halide on the partial positive charge on the carbon of the methyl halides.

A Because the halides decrease in electronegativity from F to I (see Table 6.2), we can assume that the amount of electron density removed from the carbon is greatest for F and least for I. This means that the carbon in methyl fluoride should be the most electrophilic and that the rate of reaction of hydroxide with the methyl halides to form the intermediate should be greatest for methyl fluoride.

Experimentally, the order of reactivity of the methyl halides toward nucleophiles is as follows: $CH_3I > CH_3Br > CH_3Cl > CH_3F$. This order is therefore inconsistent with formation of the trigonal bipyramidal intermediate in a rate-determining step.

REACTION OF *tert*-BUTYL CHLORIDE WITH WATER: A TWO-STEP IONIC MECHANISM

The rate of the substitution reaction of methyl chloride is also highly dependent on the nature of the nucleophile. We have already seen that hydroxide, a strong base, is also a good nucleophile. If we were to allow methyl chloride to react with the poor nucleophile water to form methanol, we would discover that this reaction is so slow at room temperature that we would not observe any chemical change in a reasonable time. It is quite remarkable, however, that if we replace the methyl chloride substrate with *tert*-butyl chloride, the reaction with water proceeds at a much faster rate. Even at room temperature this reaction is complete in a matter of minutes. What accounts for this dramatic rate difference with water when we substitute *tert*-butyl chloride for methyl chloride? Because the reaction uses water, which is a poor nucleophile, the enhanced rate must be related to the change in structure of the alkyl chloride.

Q Write an equation for the reaction of *tert*-butyl chloride with water.

A

The experimentally determined rate law for the reaction of *tert*-butyl chloride with water is

$$\text{Rate} = k[\textit{tert}\text{-butyl chloride}]$$

Unlike the reaction of methyl chloride with hydroxide ion, which is second-order overall, this reaction is first-order. The reaction of *tert*-butyl chloride with water is therefore denoted as an S_N1 reaction.

Q What does the designation S_N1 mean?

A The number 1 refers to the *molecularity* or number of molecules or ions involved in the rate-determining step of the reaction. Thus, the reaction of *tert*-butyl chloride with water has a rate law that indicates that only one molecule is involved in the rate-determining step. The reaction is therefore unimolecular. Of course, S_N refers to the category of reaction—nucleophilic substitution. ■

The observation of a different rate law is a clear indication that the reaction of *tert*-butyl chloride with water occurs by a different mechanism, one in which only *tert*-butyl chloride is involved in the rate-determining step. Clearly, this new mechanism must have more than one step because only *tert*-butyl chloride participates in the slow step. Because the rate law is first-order, the rate-limiting step is unimolecular.

Previously we considered a mechanism in which a C–Cl bond breaks heterolytically in a first step to yield a carbocation and a chloride ion. If this were to occur in the reaction between water and *tert*-butyl chloride to form *tert*-butyl alcohol and aqueous hydrogen chloride, the intermediate cation would be the trigonal planar *tert*-butyl carbocation. Although we cannot provide a comprehensive justification for the stability of this tertiary carbocation, three methyl groups attached to the electron-deficient carbon in the carbocation can assist in stabilizing the positive charge. Note in the following S_N1 mechanism that we have labeled the first step as the rate-determining step (RDS).

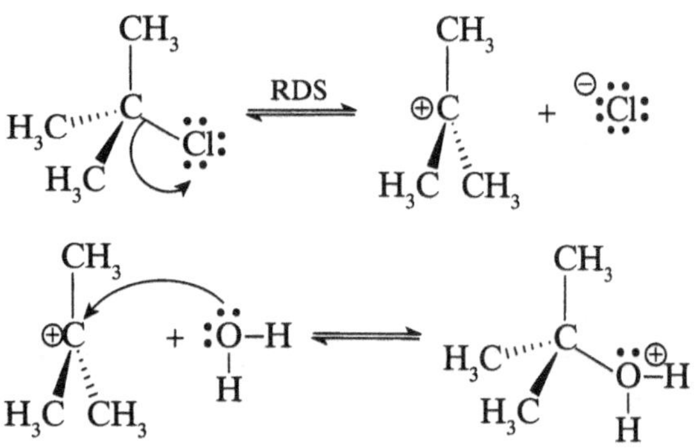

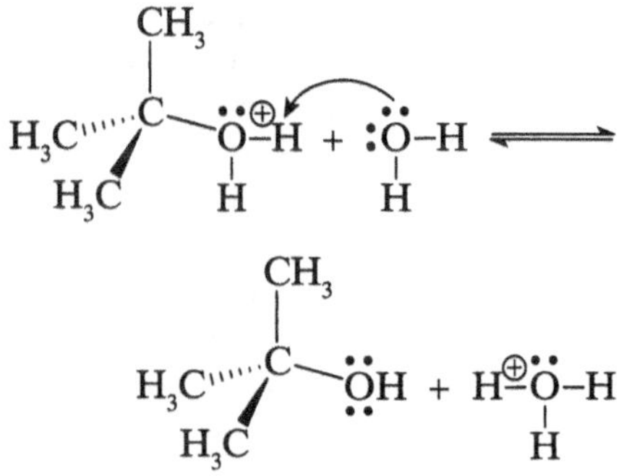

In the second step, the attack of the water molecule on the carbocation results in a positively charged intermediate. This intermediate is subsequently deprotonated by another water molecule from the solution to yield *tert*-butyl alcohol and a hydronium ion.

Q Could the chloride ion, produced in the first step, deprotonate the protonated *tert*-butyl alcohol in the third step?

A The chloride ion is the conjugate base of a strong acid (HCl) and is therefore a very weak base. In other words, the reaction

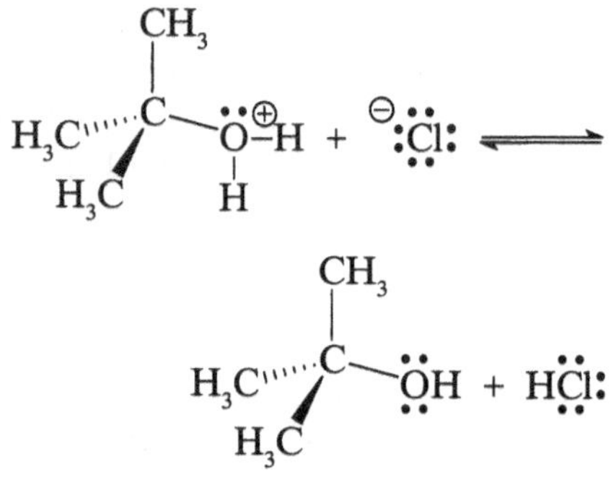

has a lower extent than the reaction of the intermediate with water (a better base):

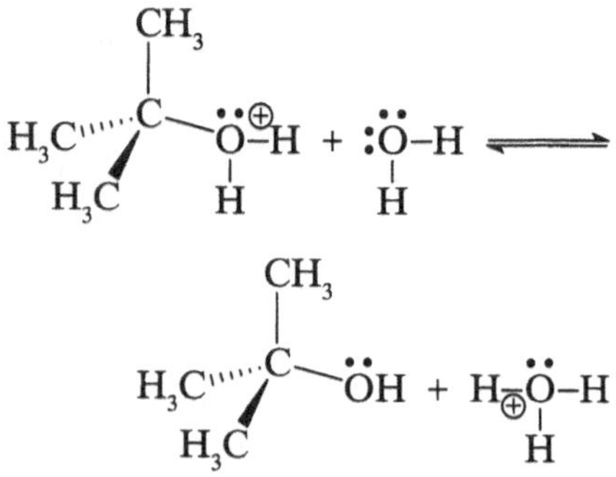

■

Because the carbocation is planar, the water could make its nucleophilic attack from either side of the cation. This mirror-image attack could not produce an enantiomeric product because the central carbon is achiral (i.e., it does not have four different substitutents).

It is noteworthy that the rate of the S_N1 mechanism is strongly dependent on solvent polarity. Changing the solvent from ethanol to water increases the rate of reaction by a factor of 10^5. The more polar water solvent is more effective in stabilizing the carbocation intermediate than is the somewhat less polar ethanol solvent. This solvent effect is consistent with the formation of a charged intermediate from a neutral reactant in the rate-determining step.

Q Write a reaction profile for the S_N1 mechanism.

A Refer to Figure 6.4.

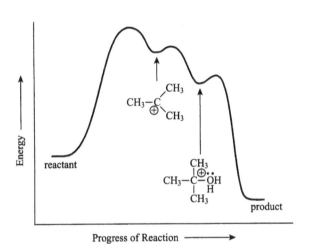

Figure 6.4. A reaction profile for the S_N1 mechanism for the reaction of *tert*-butyl chloride with water.

The intermediates shown in the reaction profile are preceded and followed by a transition state.

This introduction to the S_N1 and S_N2 reaction mechanisms has emphasized that a number of variables influence the rates of these reactions. The structure of the substrate, the nature of the nucleophile and leaving group, and the characteristics of the solvent are all important factors. Whereas methyl halides and primary alkyl halides (e.g., ethyl bromide) generally undergo substitution via an S_N2 mechanism, tertiary substrates, such as *tert*-butyl chloride, undergo substitutions by an S_N1 mechanism.

Q Predict the product and mechanism of the reaction of methoxide ion with bromomethane.

A

$$CH_3Br + CH_3O^- \rightarrow CH_3OCH_3 + Br^-$$

The mechanism can be proposed to be S_N2 because the substrate is methyl bromide and methoxide is a potent nucleophile.

Q Explain why benzyl chloride participates via an S_N2 reaction with hydroxide ion but via an S_N1 reaction with water.

A As a primary alkyl halide the benzyl chloride may be readily attacked by a reactive nucleophile such as the hydroxide ion.

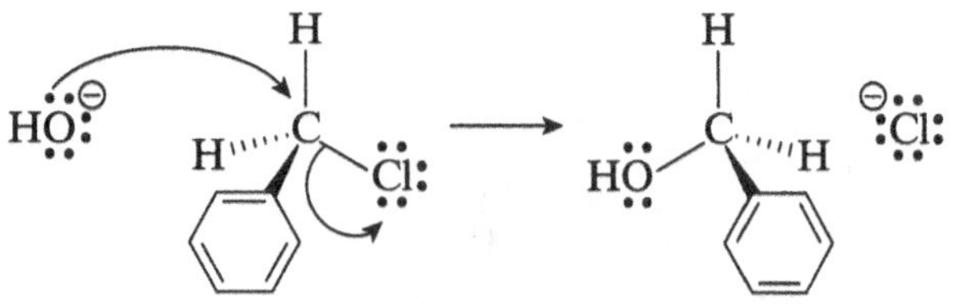

In the presence of water, a polar protic solvent that will stabilize a cation and the chloride ion, the C–Cl bond can break heterolytically to form a carbocation that is resonance-stabilized by the phenyl group. This cation is subsequently attacked by water just as the *tert*-butyl cation was attacked in the example given in the text.

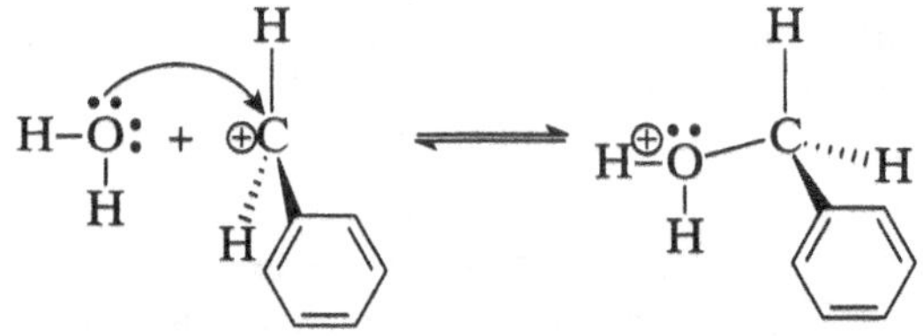

Resonance stabilization of the benzyl cation (the $C_6H_5CH_2-$ group is called the *benzyl group*)s is shown below:

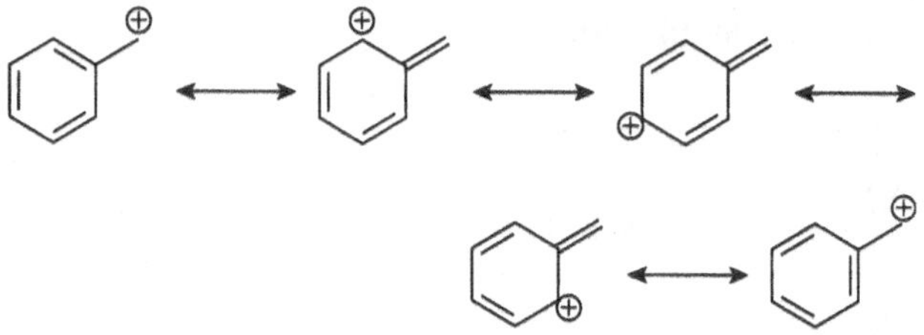

INDEX

Absolute configuration, 60–62
Acid anhydrides, 28
Acid derivatives, 27
Acid halides, 28
Activation energy, 66–67, 69, 91
Addition reaction, 87
Alcohols, 21–23
Aldehydes, 25–26
Alkanes, 10
Alkanyl names, 15
Alkenes, 16
Alkynes, 18
Amides, 28
Amines, 29
Amino acids, 30, 58
Amphiprotic, 75
Aprotic solvent, 97–98
Arenes, 19
Aromatic compounds, 9, 19
Arrhenius equation, 68–69

Backside attack, 93
Basicity, 84
Benzene, 1–5, 18–20, 39–40, 63
Boltzmann energy distribution, 67, 69
Bond cleavage, 88
Bond energies, 73–74
Bond order, 37
Branched structural formula, 10–15
Bronsted–Lowry reaction, 65, 71–79
    effect of structure, 75–79
    extent, 76
    inductive effect, 77
    resonance effect, 77
    steric effect, 79

Cahn–Ingold–Prelog rules, 60
Carbanion, 50
Carbocation, 50
Carboxylic acids, 26–27
Cis and *trans* isomers, 17
Collision frequency, 69
Condensed formulas, 6
Concerted mechanism, 88, 93, 95, 97
Configuration, 60
Connectivity, 2, 51
Chirality, 57
Conjugate acid/base, 75–78
Cycloalkanes, 16

Dextrorotatory, 59
Delocalization of electrons, 42
Diastereomers, 58
Dipole–dipole interaction, 97

Electron configuration, 34
Electron pushing, 36, 40, 67, 74
Electron-sharing reaction, 80
Electrophile, 83
Elimination reaction, 87
Empirical formula, 1
Enantiomers, 57
Endothermic, 69, 72
Enthalpy, 71
    of formation, 73
Entropy, 71
Equilibrium constant, 64
Esters, 27–28
Ethers, 23–24
Exothermic, 69, 72
Extent of reaction, 63, 71
E,Z-system for alkenes, 18

Fischer projections, 62
Formal charge, 38

The Bridge to Organic Chemistry: Concepts and Nomenclature
By Claude H. Yoder, Phyllis A. Leber, and Marcus W. Thomsen
Copyright © 2010 John Wiley & Sons, Inc.

INDEX

Front-side attack, 93–94
Functional groups, 9, 21

Geometric isomers, 17, 55–57
Gibbs energy, 71
  of formation, 73

Heterolytic bond cleavage, 65, 66, 88
Homolytic bond cleavage, 65, 66, 88
Hybrid orbitals, 46
Hydrocarbons, 9
Hydrocarbon substituents, 11

Inductive effect, 77
Intermediate, 66, 95, 99
Intermolecular forces, 55
Ion–dipole interaction, 66, 97
Isoelectronic, 42
IUPAC rules, 9, 12, 14–16, 18, 24
Isomerism, 51
Isomers, classification of, 58

Ketones, 24–26

Leaving group, 93, 95
Levorotatory, 59
Lewis acid–base reaction, 80
Lewis model, 33–40
Line formulas, 5, 6
Lewis structures, 2–4, 33–40
Lone pair electrons, 35

Mechanism, 65, 87
  $H_2 + Cl_2$, 88
  $CH_4 + Cl_2$, 89
  $CH_3Cl + OH^-$, 92
  $(CH_3)_3CCl + OH^-$, 99–102
  evidence for, 88–102
  ionic, 92
  radical, 88
Meso compound, 60
Meta substituent, 20
Molecular formula, 2
Molecularity, 100

Nitriles, 28
Nonbonded electrons, 35
Nucleophile, 83
Nucleophilicity, 84

Octet rule, 3, 11, 34, 43
  exception to, 43
One-step mechanism, 88, 93, 95 97, 98
Optical isomerism, 57–62
Orbitals, 44
Order of reaction, 68
Ortho substituent, 20

Para substituent, 20
Percent composition, 1

Percent yield, 64
Phenols, 23
Polarimeter, 59
Potential energy diagram, 94, 96, 98
Primary substituent, 15
Propagation step, 89
Protic solvent, 97–98, 102
Proton transfer reactions, 74
  in organic, 79

R,S designations, 60
Racemic mixture, 59
Radical, 65, 88
Rate constant, 68
Rate determining step, 90–93, 98–101
Rate laws, 68
Rate of reaction, 63, 67
Rearrangement reaction, 87
Reaction profile, 66, 91, 96, 97, 101
Resonance, 37
Resonance effect, 78
Resonane energy, 42
Resonance forms, 37
Resonance structures, contributors, 37
Resonance hybridization, 4
Rotation, lack of, around C=C, 47

Secondary substituent, 15
S_N1 mechanism, 99–101
S_N2 mechanism, 98
Solvent effect, 97
Spectroscopic methods, 3
Stereochemistry, reactions and, 93
Stereoisomerism, 55
Steric effect, 79
Structural formula, 2, 5
  drawing conventions, 5
Structural isomer, 51–54
Substituent, hydrocarbon, 11, 14
Substituent, other, 14
Substitution reaction, 87
Substrate, 93–94

Tertiary substituent, 15
Thermodynamics, 71
Trans isomer, 17
Two-step mechanism, 99
Transition state, 88, 95–99

Unsaturated compounds, 9

Valence bond model, 44–48
Valence shell, 34
VSEPR model, 49

Yield, 64